Letts

GCSE EXAM
SECRETS

CHEMISTRY

GCSE
Exam
Secrets

Chemistry

Bob McDuell

CONTENTS

To revise any of these topics more thoroughly, see *Letts Revise GCSE Chemistry Study Guide.*

(see inside back cover for how to order)

THIS BOOK AND YOUR GCSE EXAMS

Introduction

This book is designed to help you get better results.

▶ Attempt the sample GCSE question on the first page of each chapter.

▶ Check your answers against the grade A and C candidates' answers and see if you could have done better.

▶ Try the exam practice questions and then look at the answers and examiner's tips.

▶ Make sure you understand why the answers given are correct.

▶ When you feel ready, try the GCSE mock exam papers.

If you perform well on the questions in this book you should do well in the examination. Remember that success in examinations is about hard work, not luck.

What examiners look for

▶ Examiners are obviously looking for the right answer, however it does not have to match the wording in the examiner's marking scheme exactly.

▶ Your answer will be marked correct if the chemistry is correct even if it is not expressed exactly as it is on the mark scheme. The examiner has to use professional judgement to interpret your answers. You do not get extra marks for writing a lot of irrelevant words.

▶ You should make sure that your answer is clear, easy to read and concise.

▶ You must make sure that your diagrams are neatly drawn and labelled clearly. You do not need to use a ruler to draw diagrams. Make sure diagrams are large enough – they are often drawn too small for the examiner to see them clearly.

▶ On many papers you will have to draw a graph. At Foundation tier, axes of the graph and scales are usually given. However, at Higher tier you usually have to choose them. Make sure you always use over half the grid given and make sure you label the axes clearly. If the graph gives points that lie on a straight line use a ruler and a sharp pencil for this line. If it is curve, draw a curve rather than join the points with a straight line. Remember you may have some anomalous points and your line or curve should not go through these.

Exam technique

▶ You should spend the first few minutes of the assessment reading through the whole question paper.

▶ Use the mark allocation to guide you on how many points you need to make and how much to write.

▶ You should aim to use one minute for each mark; thus if a question has 5 marks, it should take you 5 minutes to answer the question.

▶ Plan your answers; do not write down the first thing that comes into your head. Planning is absolutely necessary in questions requiring continuous and extended answers.

▶ Do not plan to have time left over at the end. If you do, use it usefully: check you have answered all the questions, check arithmetic and read longer answers to make sure you have not made silly mistakes or missed things out.

DIFFERENT TYPES OF QUESTIONS

Questions on end of course examinations (called terminal examinations) are generally structured questions. Approximately 25% of the total mark, however, has to be awarded for answers requiring extended and continuous answers.

Structured questions

A structured question consists of an introduction, sometimes with a table or a diagram followed by three to six parts, each of which may be further sub-divided. The introduction provides much of the information to be used, and indicates clearly what the question is about.

▸ Make sure you read and understand the introduction before you tackle the question.

▸ Keep referring back to the introduction for clues to the answers to the questions.

Remember the examiner only provides the information that is required. For example, if you are not using data from a table in your answer you are probably not answering it correctly.

Structured questions usually start with easy parts and get harder as you go through the question. Also you do not have to get each part right before you tackle the next part.

Some questions involve calculations. Where you attempt a calculation you should always include your working. Then if you make a mistake the examiner might still be able to give you some credit. It is also important to include units with your answer.

Some structured questions require parts to be answered with longer answers. A question requiring continuous writing needs two sentences with linked ideas. A question requiring extended writing may require an answer of six to ten lines. Candidates taking GCSE examinations generally do less well at questions requiring extended and continuous writing. They often fail to include enough relevant scoring points and often get them in the wrong order.

Marks are available on the paper for Quality of Written Communication (QWC). To score these, you need to write in correct sentences, use scientific terminology correctly and sequence the points in your answer correctly.

 This logo in a question shows that a mark is awarded for QWC.

Approximately 5% of the marks are awarded for questions testing Ideas and Evidence. Usually these questions do not require you to recall knowledge. You have to use the information given to you in the question.

WHAT MAKES AN A/A*, B OR C GRADE CANDIDATE

Obviously, you want to get the highest grade that you possibly can. The way to do this is to make sure that you have a good all-round knowledge and understanding of Chemistry.

GRADE A* ANSWER

The specification identifies what an A, C and F candidate can do in general terms. Examiners have to interpret these criteria when they fix grade boundaries. Boundaries are not a fixed mark every year and there is not a fixed percentage who achieve a grade each year. Boundaries are fixed by looking at candidates' work and comparing the standards with candidates of previous years. If the paper is harder than usual the boundary mark will go down. The A* boundary has no criteria but is fixed initially as the same mark above the A boundary as the B is below it.

GRADE A ANSWER

A grade candidates have a wide knowledge of Chemistry and can apply that knowledge to novel situations. An A grade candidate generally has no bad questions and scores marks throughout. An A grade candidate has to have sat Higher tier papers. The minimum percentage for an A grade candidate is about 80%.

GRADE B ANSWER

B grade candidates will have a reasonable knowledge of the topics identified as Higher tier only. The minimum percentage for a B candidate is exactly halfway between the minimum for A and C (on Higher tier).

GRADE C ANSWER

C grade candidates can get their grade either by taking Higher tier papers or by taking Foundation tier papers. There are some questions common to both papers and these are aimed at C and D candidates. The minimum percentage for a C on Foundation tier is approximately 65% but on Higher tier it is approximately 45%.

If you are likely to get a grade C or D on Higher tier you would be seriously advised to take Foundation tier papers. You will find it easier to get your C on Foundation tier as you will not have to answer the questions targeted at A and B.

HOW TO BOOST YOUR GRADE

Grade booster ┈┈┤> How to turn C into B

▶ All marks have the same value. Don't forget the easy marks are just as important as the hard ones. Learn the definitions – these are easy marks in exams and they reward effort and good preparation. If you want to boost your grade, you **cannot** afford to miss out on these marks – they are easier to get.

▶ Look carefully at the command word at the start of the sentence. Make sure you understand what is required when the word is **State, Suggest, Describe, Explain** etc.

▶ For numerical calculations, always include units.

▶ If the question asks for the name of a chemical, do not give a formula.

▶ Read the question twice and underline or highlight key words in the questions. *e.g. in terms of covalent bonding suggest why nitrogen is less reactive than chlorine.*

▶ Use the Periodic Table and other data that you are given. *Don't try to remember data such as relative atomic masses.*

▶ When writing the structural formula of an organic compound make sure that each carbon forms four bonds and you have not missed off hydrogen atoms.

Grade booster ┈┈┤> How to turn B into A/A*

▶ Always write balanced equations for chemical reactions. This is one of the ways chemists communicate information. Make sure you use the right symbols and formulae in your equations.

▶ Make sure that the equation you have written is correctly balanced. Wrongly balanced equations will cost you a mark each time. Ionic equations must have the same total charge on the left hand side and the right-hand side.

▶ Check any calculations you have made at least twice, and make sure that your answer is sensible. For example, if you divide 0.49 by 1.9 the answer should be approximately 0.25.

▶ Make sure you know the difference between number of decimal places and number of significant figures. *25.696 to one decimal place is 25.7 and to 4 significant figures it is 25.70*

▶ Give complete colour changes. The test for an alkene is **NOT** that bromine turns colourless, but *the colour change is from brown to colourless.*

▶ Try to give conditions for chemical reactions, e.g. concentrated or dilute acid or heat to 140°C.

▶ In questions requiring extended writing make sure you make enough good points and you don't miss out important points. Read the answer through and correct any spelling, punctuation and grammar mistakes.

CHAPTER 1

Classifying materials

To revise this topic more thoroughly, see Chapter 1 in *Letts Revise GCSE Chemistry Study Guide*.

> Try this sample GCSE question and then compare your answers with the Grade C and Grade A model answers on the next page.

There are two different types of lithium atom, $^{6}_{3}$Li and $^{7}_{3}$Li.

a What name is given to these different types of atom?

Isotope .... **[1]**

b Describe the structures of the two atoms of lithium.

Use your knowledge of particles in atoms in your answer.

Both lithium has 3 protons & 3 electron. However $^{7}_{3}$Li
has one more electron/neutron than $^{6}_{3}$Li, 9e. $^{7}_{3}$Li has
4 neutrons & $^{6}_{3}$Li has 3 neutrons .... **[4]+[1]**

c The relative atomic mass of lithium is 6.9.

What information does this give about the proportions of these two types of atom in a sample of lithium? Explain your answer.

$x\%/100 \times 6 + 1\cdot x\%/100 \times 7 = 6x/100 + 7 - 7x/100$
$x/100 = 7$ $7 - x/100 = 6.9$ $0.1 = x/100$ $x = 10$ **[2]**
10% of $^{6}_{3}$Li & 90% of $^{7}_{3}$Li

d A fluorine atom has an electron arrangement of 2,7.   *$Li^{+} F^{-}$*

Describe the changes in electron arrangement that take place when lithium and fluorine combine.

Li^{+} $1s^{2} 2s^{1}$ $\rightarrow$ $1s^{2} 2s^{2} 2s^{2}$ & $F^{-} 1s^{2} 2s^{2} 2p^{5}$ ....

....

.... **[3]**

e Finish the table showing the properties of lithium fluoride.

state at room temperature	*solid*
solubility in water	*soluble*
solubility in hexane	*insoluble*

[3]

(Total 14 marks)

These two answers are at grades C and A. Compare which one your answer is closest to and think how you could have improved it.

GRADE C ANSWER

Asif has made the mistake of confusing isotopes and isomers. This is frequently seen.

Asif

a   Isomer ✗

b   Lithium–6 contains six nucleons and lithium–7 contains 7. ✓
Both contain 7 electrons. ✓ ✗✗
 ✗

In this answer the use of the term nucleon, while correct, does not differentiate protons and neutrons. The QWC mark was not awarded because the correct use of protons, neutrons and electrons was required.

Only one mark as no explanation is given.

c   More lithium–7 than lithium–6. ✓ ✗

d   Lithium atom 2.1. ✓
One electron is transferred between the atoms. ✗✗

e   solid ✓ soluble ✓ insoluble ✓

Here it is not clear whether electrons are transferred from lithium to fluorine or fluorine to lithium.

7 marks = Grade C Answer

Grade booster ···} move a C to a B
It is in **b** and **c** that Asif has most opportunities to improve. His answer lacks detail.

GRADE A ANSWER

This is a very good answer and the only errors are in **e** where the properties of ionic compounds have to be recalled.

Becky

a   Isotope ✓

b   Both lithium atoms contain 3 protons ✓ and 3 electrons. ✓
Lithium–6 contains 3 neutrons. ✓
Lithium–7 contains 4 neutrons. ✓
 ✓

Correct use of terms protons, neutrons and electrons.

c   More lithium–7 than lithium–6. ✓
RAM is much closer to 7 than 6. ✓

d   Electron arrangement of lithium 2,1. ✓
Each lithium atom loses 1 electron. ✓
Each fluorine atom gains 1 electron. ✓

e   solid ✓ insoluble ✗ soluble ✗

This is a very creditable answer.

12 marks = Grade A answer

Grade booster ···} move A to A*
Becky's errors are in **e**. This is not the hardest part of the question. Remember all marks are of equal importance. There is no benefit getting hard parts right if you get easier parts incorrect.

Calcium and oxygen react together to form calcium oxide.

a Write a symbol equation for this reaction.

$2Ca_{(s)} + O_{2(g)} \rightarrow 2CaO_{(s)}$ [3]

b (i) What type of bonding is present in calcium oxide?

Ionic [1]

(ii) Draw diagrams to show the electron arrangements in a calcium and an oxygen atom.

Calcium atom

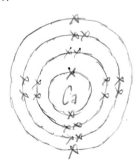

Oxygen atom

[2]

(iii) Describe the changes in electron arrangement when calcium and oxygen combine.

Ca transfer 2 electrons to Oxygen
ie Ca^{++} $1s^2 2s^2 2p^6 3s^2$ $\rightarrow$ $1s^2 2s^2 2p^6$
      O^{--} $1s^2 2s^2 2p^2$ $\rightarrow$ $1s^2 2s^2 2p^6$ [3]

(iv) Explain why calcium oxide has a very high melting point.

Ionic bond 2^+ & 2^- electrostatic force is stronger
than 1^+ & 1^-

[3]

(Total 12 marks)

These two answers are at grades C and A. Compare which one your answer is closest to and think how you could have improved it.

GRADE C ANSWER

One mark awarded as David gave the correct formula for the product calcium oxide.

No mention of the number of electrons transferred.

David

a $Ca + O \rightarrow CaO$ ✗✓✗

b (i)   Ionic ✓

(ii)   2,8,8,2   2,6 ✓

(iii)   Calcium loses electrons, oxygen gains. ✓✓✗

(iv)   Ions held together by electrostatic forces. ✓

David failed to draw the diagrams asked for in the question. The examiner awarded 1 mark because the electron arrangements given were correct.

The answer did not refer to calcium oxide but to any ionic solid therefore only one mark awarded.

6 marks = Grade C Answer

Grade booster ⋯⟩ move a C to a B
David should make sure he answers the question exactly as set and includes all detail.

GRADE A ANSWER

This is an excellent answer scoring all marks well in a and b.

Sam

a $2Ca + O_2 \rightarrow 2CaO$ ✓✓✓

b (i)   Ionic ✓

(ii)

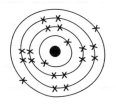

✓✓

(iii)   Each calcium atom loses two electrons, each oxygen atom gains two electrons. ✓✓✓

c   Ions Ca^{2+} and O^{2-} ✓

Strong electrostatic forces ✓✗

The answer does not explain why calcium oxide has a very high melting point. This is due to the electrostatic charge between 2+ and 2− ions which will be greater than between 1+ and 1− ions.

11 marks = Grade A Answer

Grade booster ⋯⟩ move A to A*
This is an excellent answer but only two points are made in **c** when three marks are available.

Classifying materials

11

Classifying materials

1 The table shows the atomic structure of six particles, represented by the letters A–F. The symbols are not the symbols of the elements.

Particle	Electrons	Protons	Neutrons
A	6	6	6
B	2	2	2
C	12	12	12
D	10	12	12
E	6	6	8
F	10	13	14

Use the letters A–F to answer the following questions.

a) Which two particles are an atom and an ion of the same element?

............*C*............ and ....*D*............ ①

b) Which two particles are positive ions?

............*D*............ and ....*F*............ ①

c) Which particle is a noble gas?

............*B*............ ①

d) Which particle has 14 nucleons?

............*E*............ ①

e) Which two particles are isotopes of the same element?

............*A*............ and............*E*............ ①

TOTAL 5

2 The diagrams show the outer shell electron arrangements in atoms of hydrogen and chlorine.

hydrogen atom

(not to scale)

chlorine atom

a) Draw diagrams to show the arrangement of outer shell electrons in

i) a hydrogen molecule, H_2

H ⊙ H

ii) a chlorine molecule, Cl_2.

Cl ⊙ Cl

②

b) What type of bonding is present in both a hydrogen molecule and a chlorine molecule?

.... *covalent* .... ①

c) Explain why the boiling point of chlorine is greater than the boiling point of fluorine, F_2, but less than the boiling point of bromine, Br_2.

....

....

.... ③

d) Dry hydrogen chloride, HCl, is a gas. It dissolves in water, producing heat energy and a solution which conducts electricity.

Describe the changes in bonding when dry hydrogen chloride is dissolved in water.

....

....

.... ③

TOTAL 9

1

3 The table gives some properties of six substances A–F.

Substance	Density (g/cm^3)	Melting point (°C)	Boiling point (°C)	Electrical conductivity	Other properties
A	0.97	89	883	conductor	forms a basic oxide
B	3.51	>3550	4827	non-conductor	forms an acidic oxide
C	2.25	3680	4827	conductor	forms an acidic oxide
D	3.12	−7	59	non-conductor	does not burn
E	0.001	−169	−104	non-conductor	burns to form CO_2 and H_2O
F	2.17	801	1413	non-conductor when solid; conductor when molten	dissolves in water

a) Which substance is a metal? ....
   Give a reason for your answer.

   .... ②

b) Which substance has a structure of widely spaced molecules?
   Give a reason for your answer.

   .... ②

c) Which substance has a giant ionic structure? ....
   Give a reason for your answer.

   .... ②

d) B and C are different forms of the same element in the same physical state.

   i) What name is given to the different forms of the same element in the same
      physical state?

      .... ①

   ii) What evidence is there in the table that B and C are the same element?

      .... ①

   iii) What evidence is there that B and C are non-metallic?

      .... ①

   TOTAL 9

4 Silicon(IV) oxide, SiO_2, has a giant structure of atoms.
   Sodium chloride, NaCl, has a giant structure of ions.
   Iodine, I_2, has a molecular structure.

a) Describe a test which you could use to decide whether a substance has a giant structure or a
   molecular structure.

   ....

   ....

   .... ③

b) Describe a test which you could use to decide whether a substance with a giant structure is made up of atoms or ions.

....

....

.... ③

TOTAL 6

5 The diagram shows the arrangement of outer electrons in a molecule of ethane, C_2H_6.

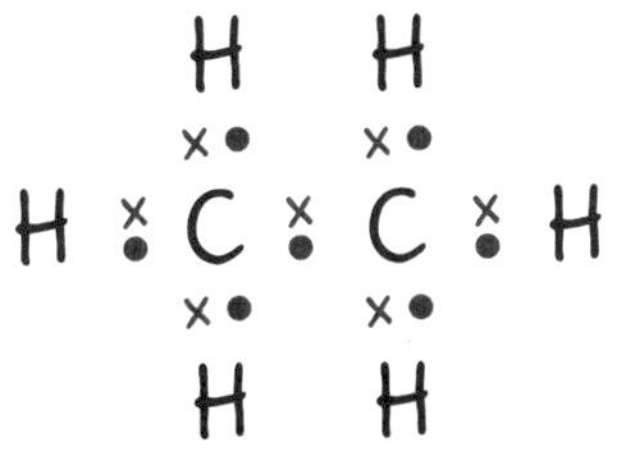

H   H
x•   x•
H x C x C x H
x•   x•
H   H

a) Suggest two physical properties of ethane.

....

.... ②

b) Draw similar diagrams to show the arrangement of outer electrons in

   i) oxygen, O_2

   ii) nitrogen, N_2

   iii) ammonia, NH_3.

③

TOTAL 5

ANSWERS ON PAGE 16 ANSWERS ON PAGE 16 ANSWERS ON PAGE 16 ANSWERS ON PAGE 16

Classifying materials

① a)   C and D   (both required) **❶**
   b)   D and F   (both required) **❶**
   c)   B   **❶**
   d)   E   **❶**
   e)   A and E   (both required) **❶**

EXAMINER'S TIP

Nucleons are particles in the nucleus – either protons or neutrons.
In a) C and D have the same number of protons and are therefore the same element but have different number of electrons.
In b) D and F have more protons than electrons.
In c) helium has two electrons.
In d) E has 6 protons and 8 neutrons.
In e) A and E contain the same number of protons and electrons and are therefore the same element but have a different number of neutrons.

② a) i) hydrogen molecule

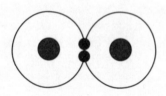

ii) chlorine molecule

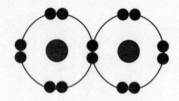

                                                              ❷

   b)   Covalent   **❶**
   c)   Fluorine, chlorine and bromine
        molecules are similar.   **❶**
        Difference in mass: bromine heaviest,
        fluorine lightest.   **❶**
        Heavier molecules need more energy
        to overcome the attractive forces between
        them so they can separate and escape to
        form a gas.   **❶**

   d)   Hydrogen chloride molecules (dry)
        contain a covalent bond.   **❶**
        On dissolving in water, the hydrogen
        chloride ionises or forms ions.   **❶**
        Ions are H^+ and Cl^- or a better answer
        would be to show the electron arrangement
        in the ions.   **❶**

EXAMINER'S TIP

You will frequently have to draw simple diagrams to represent atoms. In many cases candidates draw them so small that detail cannot be seen by the examiner. In these diagrams it is usual only to show the nucleus and the outer electrons. If you attempt to show inner electrons incorrectly, you may lose marks. This question tests your understanding of covalent and ionic bonding. The change in the bonding of hydrogen chloride takes place when water is present.

③ a) i) A   **❶**
        Forms a basic oxide   **❶**
   b)   E   **❶**
        Has the lowest density/is a gas   **❶**
   c)   F   **❶**
        Conductor of electricity when molten
        and dissolves in water   **❶**
   d) i) Allotropes   **❶**
     ii) They have the same boiling point   **❶**
    iii) They form acidic oxides   **❶**

EXAMINER'S TIP

This question is essentially a data handling question where you have to apply your knowledge.

④ a)   Heat each substance.   **❶**
        Substance with a giant structure will
        have a high melting point.   **❶**
        Substance with a molecular structure
        will have a low melting point.   **❶**
   b)   Test the electrical conductivity of a
        substance when it is molten.   **❶**
        A substance with a giant structure of
        ions will conduct electricity when
        molten.   **❶**
        A substance with a ~~giant structure~~ molecular structure
        of atoms will not.   **❶**

FOR MORE INFORMATION ON THIS TOPIC ... SEE REVISE GCSE CHEMISTRY ... CHAPTER 1

5 a) Any two from:
gas
insoluble in water
soluble in organic solvents ❷

b) i)

ii)

iii)

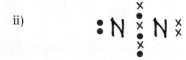

❸

CHAPTER 2

Changing materials

To revise this topic more thoroughly, see Chapter 2 in *Letts Revise GCSE Chemistry Study Guide*.

 Try this sample GCSE question and then compare your answers with the Grade C and Grade A model answers on the next page.

Aluminium is found in many parts of the world as bauxite. This is purified at the mining site to produce pure aluminium oxide.

a   Suggest one advantage to the mining company of purifying bauxite at the mining site.

....

.... **[1]**

Aluminium is extracted from the pure aluminium oxide, Al_2O_3, in a smelter at 1000°C.

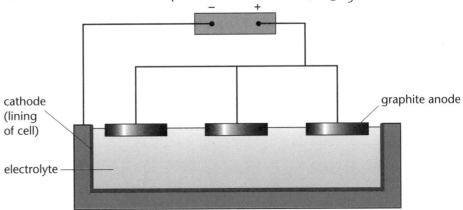

cathode (lining of cell)

graphite anode

electrolyte

b   Explain the changes that take place at the anode and cathode.

 Use ideas of oxidation and reduction in your answer.

Anode...

....

....

....

Cathode...

....

....

.... **[6]+[1]**

c   Suggest **two** reasons why the extraction of aluminium is an expensive process.

(i)   ....

(ii)   .... **[2]**

(Total 10 marks)

These two answers are at grades C and A. Compare which one your answer is closest to and think how you could have improved it.

GRADE C ANSWER

Lucy

a   Reduces transport costs ✓

b   Oxygen produced ✓ anodes burn in oxygen ✓ forming carbon dioxide ✓ Aluminium produced ✓

c   A lot of energy is required. ✓ ✗

This answer is a little vague and only scores 1 out of 2.

This answer does not refer to oxidation and reduction. One mark for QWC is for the correct use of the terms oxidation and reduction.

6 marks = Grade C Answer

Grade booster ┄┄> move a C to a B

The performance on the question is going to depend upon the marks scored in **b** where 7 marks are available. Lucy should concentrate on the parts of the question where a lot of marks are allocated.

GRADE A ANSWER

Amy

a   Create jobs locally ✗

b   Oxide ions lose electrons ✓ forming oxygen ✓ aluminium ions gain electrons ✓ forming aluminium ✓ Oxidation at anode/reduction at cathode ✓ – oxidation involves loss of electrons ✓✓

c   High temperature required to melt cryolite ✓ Large quantity of electricity used ✓

This answer is probably correct but the benefit is not to the mining company.

Very good answers

9 marks = Grade A Answer

Grade booster ┄┄> move A to A*

Amy's answer was very good. Only in **a** was there room for improvement. Here Amy did not answer the question as set.

Cycloalkanes are a family of hydrocarbons.

The simplest members of the family are shown below.

cyclopropane   cyclobutane   cyclopentane

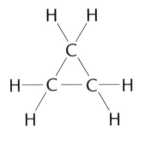

a   Why are these hydrocarbons saturated?

.... **[1]**

b   Write the name and molecular formula of the cycloalkane containing six carbon atoms.

.... **[2]**

c   Propene has the same molecular formula as cyclopropane.

   (i) Draw the structural formula of propene.

                                                                                                    [1]

   (ii) What name is given to compounds with the same molecular formulae but different structural formulae?

.... **[1]**

   (iii) Describe a test that could be used to distinguish propene and cyclopropane.

....

.... **[3]**

d (i) Write a balanced symbol equation for the burning of cyclopropane in a plentiful supply of air.

.... **[3]**

   (ii) How would the products be different if the amount of air available was restricted?

.... **[1]**

                                                                    (Total 12 marks)

GRADE C ANSWER

Chris

The formula given by Chris is true for alkanes but not the cycloalkanes in the question. The important thing about saturated hydrocarbons is the single carbon-carbon bonds throughout.

a   All fit a formula C_nH_{2n+2} ✗

b (i)   Cyclohexane ✓ C_6H_{14} ✗

c (i)   ✗

Cyclohexanes fit the same general formula as alkenes.

This structure is wrong as one carbon has five bonds and another has three.

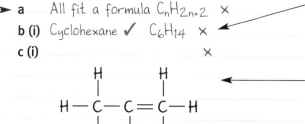

Chris should have stated that there would be no change with cyclohexane.

The equation is balanced but most people would multiply through by 2 to get whole numbers.

The question does not ask for the name of the product so the examiner was prepared to accept the correct formula.

(ii)   isotopes ✗

(iii)   Bromine solution ✓
        Propene turns colourless ✓

d (i) $C_3H_6 + \frac{9}{2}O_2 \rightarrow 3CO_2 + 3H_2O$ ✓✓✓

(ii)   CO produced ✓

Chris has confused isotopes and isomers.

7 marks = Grade C Answer

Grade booster ⋯⟩ move a C to a B
There are a number of errors in Chris's answer that could be avoided. At least two extra marks could be scored. Don't lose the easy marks. All marks have the same value.

GRADE A ANSWER

Dinah

a   All carbon-carbon bonds are single bonds ✓

b   Cyclohexane ✓ C_6H_{12} ✓

c (i)   ✓

(ii)   isomers ✓

The solution is clear throughout. It is the change of colour that must be commented upon.

(iii)   bromine water ✓
        propene goes clear ✗
        no change with cyclopropane ✓

d (i) $C_3H_6 + 9O \rightarrow 3CO_2 + 3H_2O$ ✗ ✓✓

(iii)   Carbon monoxide formed ✓

Dinah has one incorrect reactant but is awarded two marks for the correct products.

10 marks = Grade A Answer

Grade booster ⋯⟩ move A to A*
A very good answer with just a couple of points where Dinah could have done better. Always take care with equations.

Changing materials

Changing materials

1 The table gives some information about some polymers and the monomers used to make them.

a) Complete the table.

Monomer	Polymer	Structure of monomer	Structure of polymer
ethene	poly(ethene)	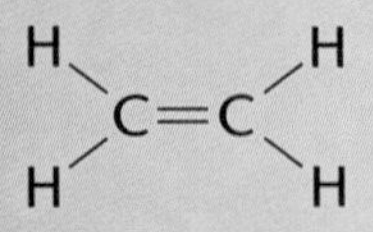 H — C = C — H (with H, H)	
chloroethene (or vinyl chloride)			
phenylethene (or styrene)			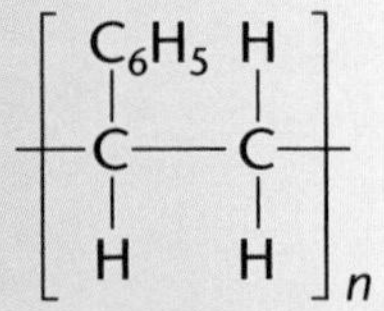 $\left[\begin{array}{cc} C_6H_5 & H \\ C & C \\ H & H \end{array}\right]_n$
	poly(propene)		

(9)

b) i) Suggest one use for the polymer made from chloroethene.

.... (1)

ii) Which material was used before the polymer was invented?

.... (1)

iii) Give one reason why the polymer is a better material for this use.

.... (1)

TOTAL 12

2 The diagram summarises how some important materials can be made from methane.

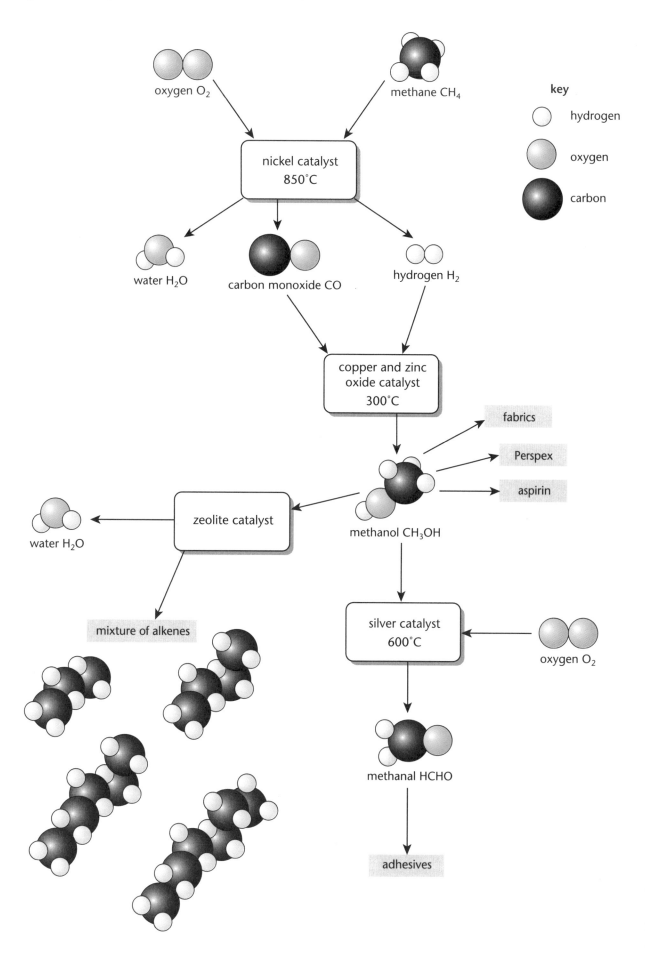

key
- hydrogen
- oxygen
- carbon

oxygen O$_2$

methane CH$_4$

nickel catalyst
850°C

water H$_2$O

carbon monoxide CO

hydrogen H$_2$

copper and zinc
oxide catalyst
300°C

fabrics

Perspex

aspirin

water H$_2$O

zeolite catalyst

methanol CH$_3$OH

mixture of alkenes

silver catalyst
600°C

oxygen O$_2$

methanal HCHO

adhesives

a) What is the important source of methane for industry?

.... ①

b) i) Write a balanced symbol equation for the reaction of methane and oxygen to produce carbon monoxide, water and hydrogen.

.... ②

ii) The catalyst in this reaction has to be heated to 850°C to start the reaction. Then heating the catalyst is not needed and the temperature remains constant.

What does this suggest about the reaction?

.... ①

c) Write a balanced equation for the reaction producing methanol.

.... ③

d) What type of reaction is the reaction of methanol to produce methanal?

.... ①

e) A mixture of alkenes can be produced when methanol is passed over a heated zeolite catalyst.

i) Suggest a use for the mixture of alkenes produced.

.... ①

ii) A zeolite structure is very open with lots of tiny holes through it.

Why does this help zeolite to be a good catalyst?

.... ①

TOTAL 10

③ A sample of antimony oxide **A** was heated strongly. It decomposes into oxygen and another oxide of antimony **B**.

1.62 g of **A** contains 1.22 g of antimony.

a) Calculate the simplest formula of **A**.

(Relative atomic mass: Sb = 122)

③

b) The relative formula mass of **A** is 324.

What is the molecular formula of **A**?

.... ①

c) The relative formula mass of **B** is 308.

What is the molecular formula of **B**?

.... ①

d) Write a balanced symbol equation for the reaction of **B** to form **A**.

.... ③

TOTAL 8

changing materials

4 Methane trapped in marine sediments as a hydrate represents such an immense carbon reservoir that it must be considered a dominant factor in estimating unconventional energy resources; the role of methane as a 'greenhouse' gas also must be carefully assessed.

This is a quotation from Dr William Dillon of the U.S. Geological Survey.

Much of the methane is trapped in ice crystals in Arctic waters.

The worldwide amounts of methane trapped in this way are conservatively estimated to total twice the amount of carbon to be found in all known fossil fuels on Earth.

Methane is one hundred times more efficient as a greenhouse gas than carbon dioxide.

a) How could methane hydrate provide a new world energy source?

.... ①

b) What effects could a rise in temperature of Arctic waters have on the conditions of the Earth's atmosphere?

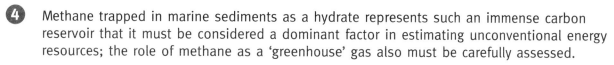

....

....

....

.... ③ + ①

TOTAL 5

5 Iron is extracted from iron ore in a blast furnace.

a) Why are metals such as magnesium and iron not found as pure metals in the Earth's crust?

.... ①

b) The diagram shows a blast furnace.

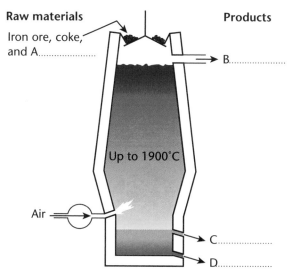

Raw materials
Iron ore, coke, and A........................

Products
B........................

Up to 1900°C

Air

C........................
D........................

Complete the diagram by labelling the missing raw material and the three missing products.

④

c) Write a symbol equation to show how carbon monoxide is formed in the furnace.

.... ③

d) The equation for the reduction reaction taking place in the furnace is shown below.

$$Fe_2O_3 + 3CO \rightarrow 2Fe + 3CO_2$$

Calculate the maximum mass of iron that could be produced from 32 tonnes of iron(III) oxide. (Relative atomic masses: Fe = 56, O = 16)

Mass = ... tonnes ③

e) Suggest two advantages and two disadvantages of building a new blast furnace close to a major city with a deep-water port.

Advantages: ....

....

Disadvantages: ....

.... ④

TOTAL 15

6 The flow diagram shows reactions involving ammonia.

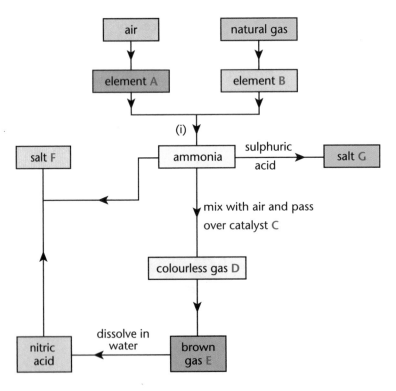

a) Write down the names of the two elements A and B used to make ammonia.

A .... B .... ②

b) i) Write down the name of the process labelled (i) in the flow diagram.

.... ①

ii) Write a symbol equation for this reaction.

.... ③

c) i) Write down the names of salts F and G.

F.. G .... ②

ii) Write down a symbol equation for a reaction producing F or G.

.... ③

iii) What is the major use of F and G?

.... ①

TOTAL 12

7 Limestone is mined in large quantities for industrial use.

a) Limestone is converted into calcium oxide in a lime kiln.

The equation for the reaction is shown below.

$$CaCO_3 \rightarrow CaO + CO_2$$

i) What type of reaction is taking place?

.... ①

ii) Calculate the mass of calcium oxide produced from 10 tonnes of calcium carbonate.
(Relative atomic masses: Ca = 40, C=12, O=16)

Mass = ... tonnes ③

iii) Finish the equation for the reaction that takes place when water is added to calcium oxide.

$$CaO + H_2O \rightarrow \text{...}$$ ①

b) Write down two other industrial uses of limestone.

i) ....

ii) .... ②

TOTAL 7

8 The Earth's atmosphere originally consisted of mainly hydrogen and helium.

Describe and explain the changes that took place to produce the atmosphere we have today.

....

....

....

....

....

....

.... ⑥+①

TOTAL 7

changing materials

① a)

Monomer	Polymer	Structure of monomer	Structure of polymer
ethene	poly(ethene)	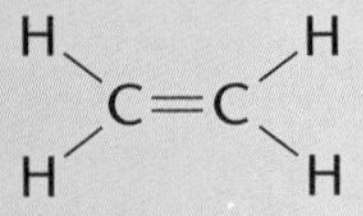	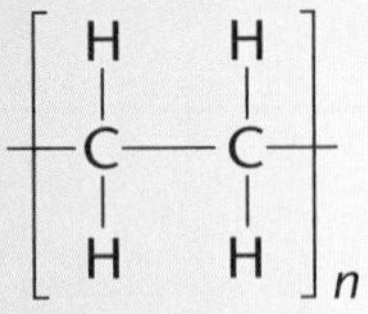
chloroethene (or vinyl chloride)	poly(chloroethene)	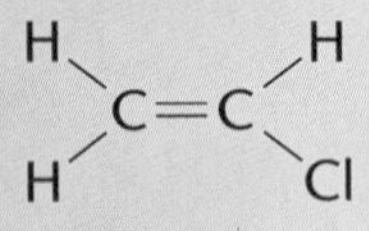	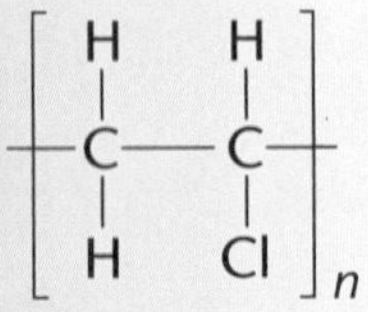
phenylethene (or styrene)	poly(phenylethene)	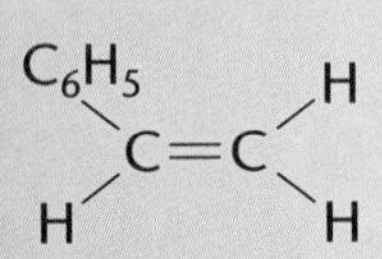	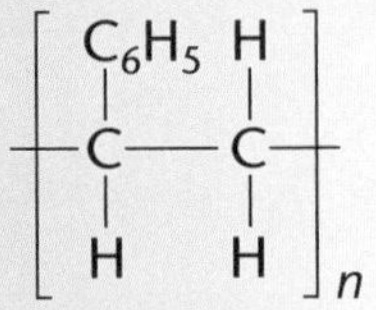
propene	poly(propene)	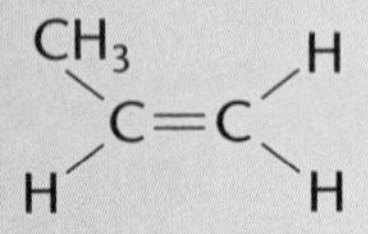	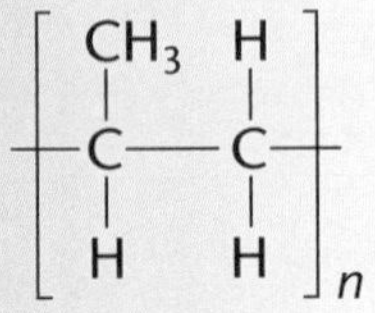

(One mark for each correct box.) **9**

b) i) Insulation for electricity cables, guttering, wellington boots. (Any one.) **1**
ii) For many of the answers above rubber was previously used. **1**
iii) Rubber perishes. It goes hard and cracks. **1**

② a) Natural gas **1**
b) i) $CH_4 + O_2 \rightarrow CO + H_2O + H_2$ **2**
(One mark for reactants and one for products.)

EXAMINER'S TIP

Candidates here will often write the combustion reaction of methane in oxygen. This is incorrect.

ii) Exothermic reaction or products have less energy than reactants. **1**

c) $CO + 2H_2 \rightarrow CH_3OH$ **3**
(One mark for reactants, one mark for products and one mark for balancing.)
d) Oxidation **1**
e) i) Making polymers (plastics) **1**
ii) Provides a large surface area for the reaction. **1**

③ a) 1.22 g of antimony combines with 0.40 g of oxygen **1**
1.22 g of Sb combines with 0.40 g of O
122   16 **1**
0.01   0.025
Simplest formula = Sb_2O_5 **1**
b) Sb_2O_5 **1**

EXAMINER'S TIP

*Work out the relative formula mass (RFM) of Sb_2O_5 (2 x 122) + (5 x 16) = 324
That confirms the simplest formula is the molecular formula.*

FOR MORE INFORMATION ON THIS TOPIC ... SEE REVISE GCSE CHEMISTRY ... CHAPTER 2

c) Sb_2O_4 ❶

EXAMINER'S TIP

The RFM of Sb_2O_4 is 308. The molecular formula must be Sb_2O_4 and not SbO_2.

d) $2Sb_2O_5 \rightarrow 2Sb_2O_4 + O_2$ ❸
(One mark for formula of reactant, one mark for formulae of products and one mark for balancing.)

❹ a) Methane is trapped within ice crystals. Melting the methane hydrate releases methane that could be used as a fuel. ❶
b) Any three points:
if temperature rises methane hydrate would melt ❶
releasing methane into the atmosphere ❶
raises the temperature of the Earth further ❶
causes more methane hydrate to melt and so on ❶
(+ 1 mark for presenting this information in a form suitable for purpose.)

EXAMINER'S TIP

You will probably know nothing about methane hydrate. You will be expected to use your knowledge of the greenhouse effect and methane to answer this question.

❺ a) These metals would react with air/water. ❶
b) A limestone ❶
B waste gases ❶
C slag ❶
D iron ❶

EXAMINER'S TIP

The layer of slag floats on top of the molten iron in the furnace because it is less dense than iron.

c) $2C + O_2 \rightarrow 2CO$ ❸
(One mark for formulae of reactants, one mark for formulae of products and one mark for balancing.)

EXAMINER'S TIP

You might write two equations here.
$C + O_2 \rightarrow CO_2$
$CO_2 + C \rightarrow 2CO$

d) RFM of iron(III) oxide =
$(56 \times 2) + (3 \times 16) = 160$ ❶
160 tonnes of Fe_2O_3 produces 112 tonnes of iron ❶
32 tonnes Fe_2O_3 produces $\underline{112} \times 32 = 22.4$ tonnes of iron   160 ❶
e) Advantages – two from:
available labour force, port for importing ore and exporting iron, possible market for iron ❷
Disadvantages – two from:
noise, pollution (air), traffic, expensive land costs ❷

❻ a) A Nitrogen ❶
B Hydrogen ❶
b)i) Haber process ❶
ii) $N_2 + 3H_2 \rightleftharpoons 2NH_3$ ❸
(One mark for formulae of reactants, one mark for formulae of products and one mark for balancing.)
c) i) F Ammonium nitrate ❶
G Ammonium sulphate ❶
ii) One equation from the following:
$NH_3 + HNO_3 \rightarrow NH_4NO_3$
$2NH_3 + H_2SO_4 \rightarrow (NH_4)_2SO_4$ ❸
(One mark for formulae of reactants, one mark for formulae of products and one mark for balancing.)
iii) Fertiliser ❶

❼ a) i) Thermal decomposition ❶
ii) RFM of $CaCO_3 = 40 + 12 + (3 \times 16) = 100$ ❶
100 tonnes $CaCO_3$ produces 56 tonnes of CaO ❶
10 tonnes of $CaCO_3$ produces 5.6 tonnes of CaO ❶
iii) $CaO + H_2O \rightarrow Ca(OH)_2$ ❶
b) Two from:
making cement
making glass
roadmaking
making iron
desulphurisation of gases from chimneys ❷

❽ Any six points from the following:
volcanoes erupted
releasing carbon dioxide and water vapour
also methane and ammonia
water condenses when Earth cooled
ammonia converted into nitrates and then into atmospheric nitrogen
methane produces carbon dioxide
oxygen produced by photosynthesis ❻
(+ 1 mark for presenting this information in a form suitable for purpose.)

changing materials

CHAPTER 3

Patterns of behaviour

To revise this topic more thoroughly, see Chapter 3 in *Letts Revise GCSE Chemistry Study Guide.*

 Try this sample GCSE question and then compare your answers with the Grade C and Grade A model answers on the next page.

The table gives information about elements in the third period of the Periodic Table.

Element	Symbol	Atomic radius (nm)	Electron arrangement
sodium	Na	0.186	2,8,1
magnesium	Mg	0.160	2,8,2
aluminium	Al	0.143	2,8,3
silicon	Si	0.117	2,8,4
phosphorus	P	0.110	2,8,5
sulphur	S	0.104	2,8,6
chlorine	Cl	0.099	2,8,7

a (i) Which element in the table is a halogen? .... **[1]**

  (ii) Which element in the table is an alkali metal? .... **[1]**

b (i) How does the atomic radius change across the period from left to right?

.... **[1]**

  (ii) Explain this pattern. Use your knowledge of atomic structure in your answer.

....

....

.... **[3]**

c What is the relationship between the electron arrangement and position of an

element in the Periodic Table?

.... **[1]**

d How would the pH values of the oxides of these elements change across the period? Explain your answer.

....

....

.... **[3]**

(Total 10 marks)

These two answers are at grades C and A. Compare which one your answer is closest to and think how you could have improved it.

GRADE C ANSWER

Potassium is an alkali metal but it is not in the table given in the question.

One vital point is missing from this answer.

Nazim

a (i) Chlorine ✓
(ii) Potassium ✗
b (i) Decreases ✓
(ii) Each time an extra proton and electron ✓
All electrons in the same shell ✓
c Group number is the same as the number of electrons ✗   *← The link is incorrect.*
d On the left-hand side oxides have pH values greater than 7 (alkaline) ✓   *← Nazim did not refer to oxides of elements in the middle of the period.*
On the right-hand side oxides have pH values less than 7 (acidic) ✓

6 marks = Grade C Answer

Grade booster ····⟩ move a C to a B
In several places Nazim lacked detail in answers. Twice two marks were scored when three marks were available. Improving these answers would boost the grade.

GRADE A ANSWER

Ben

a (i) Chlorine ✓
(ii) Sodium ✓
b (i) Decreases ✓
(ii) Each time an extra proton and electron ✓
All electrons in the same shell ✓
Increased attraction between nucleus and outer electrons ✓

Only 2 out of 3 marks was awarded. The question asks how pH values changed across the period. Ben made references to acidic, neutral and alkaline but did not mention pH.

c Group number is the same as the number of electrons in the outer shell ✓
d On the left-hand side oxides are alkaline. In the middle oxides are neutral. On the right-hand side oxides are acidic. ✓ ✓

9 marks = Grade A Answer

Grade booster ····⟩ move A to A*
This question is about the elements in a period of the Periodic Table. Often questions refer to elements in a particular group. This answer is almost perfect.

Patterns of behaviour

Sam carries out an experiment to compare the rate of reaction of calcium carbonate with dilute hydrochloric acid. She uses equal masses of large lumps and small lumps. She adds each sample of calcium carbonate lumps to 25 cm^3 of dilute hydrochloric acid to a flask.

She puts a piece of cotton wool into the neck of the flask. She measures the mass of the flask and contents at intervals. Her results are shown in the table. The equation for the reaction is shown below.

Time (min)	Total loss of mass of flask and contents	
	Large lumps (g)	Small lumps (g)
0	0.00	0.00
0.5	0.80	2.40
1	1.30	3.30
2	2.20	4.20
3	2.80	4.40
4	3.30	4.40
5	3.70	4.40
6	3.90	4.40

$$CaCO_3(s) + 2HCl(aq) \rightarrow CaCl_2(aq) + CO_2(g) + H_2O(l)$$

a   Explain why the mass of the flask and contents decreases during the reaction.

.... **[1]**

b   Why is a piece of cotton wool put into the neck of the flask?

.... **[1]**

c   Draw two graphs showing total loss of mass (on *y*-axis) against time.

Label the two graphs **large lumps** and **small lumps**.

[4]

d   Explain the shape of the graph she obtained using small lumps.

Some lumps remained at the end of the experiment.

....

....

....

.... **[4]**

(Total 10 marks)

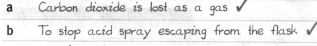

These two answers are at grades C and A. Compare which one your answer is closest to and think how you could have improved it.

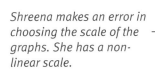**GRADE C ANSWER** Shreena

a   Carbon dioxide is lost as a gas ✓

b   To stop acid spray escaping from the flask ✓

c

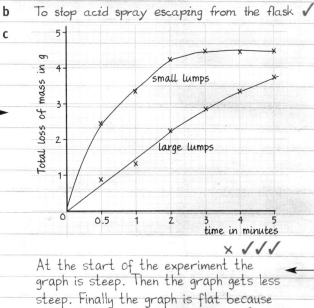

Shreena makes an error in choosing the scale of the graphs. She has a non-linear scale.

✗ ✓ ✓ ✓

At the start of the experiment the graph is steep. Then the graph gets less steep. Finally the graph is flat because all of the acid had been used up. ✓ ✗✗✗

Shreena does not answer the question set. She describes the shape of the graph rather than explains the shape.

6 marks = Grade C Answer

Grade booster ⋯⟩ move a C to a B
One of the biggest source of errors comes from candidates answering a question different from the one set. This is what Shreena does here.

GRADE A ANSWER Callum

a   Carbon dioxide is lost as a gas ✓

b   To stop carbon dioxide escaping from the flask ✗

c

This is a common error. The piece of cotton wool in the neck of the flask is not to stop gas escaping. If it was possible to stop the gas escaping, there would be no mass loss. It is there to stop acid spray escaping from the flask.

✓ ✓ ✓ ✓

d   At the start of the experiment the graph is steep as there are a lot of acid particles hitting lumps. As the acid is used up there will be fewer collisions between acid and lumps. When all acid used up the reaction has stopped and the graph is flat. ✓ ✓ ✓ ✓

9 marks = Grade A Answer

Grade booster ⋯⟩ move A to A*
This is an excellent answer. The only error is in **b**.

Patterns of behaviour

Patterns of behaviour

1 The table gives some of the properties of the elements in group 2 of the Periodic Table.

Element	Symbol	Reaction with water	Formula of chloride
Beryllium	Be		$BeCl_2$
Magnesium	Mg	Reacts slowly with cold water, vigorously with steam	$MgCl_2$
Calcium	Ca	Reacts steadily with cold water	$CaCl_2$
Strontium	Sr	Reacts vigorously with cold water	$SrCl_2$
Barium	Ba	Reacts very vigorously with cold water	

a) Arrange the four elements in italics in order of reactivity, putting the most reactive first.

.... ①

b) i) Suggest what would happen when beryllium is added to cold water.

....

.... ①

ii) Write a balanced equation for this reaction.

.... ③

c) Suggest how barium should be safely stored.

....

.... ①

d) When barium is heated in chlorine gas they react to form a solid.

i) Give the name and formula of the solid formed.

.... ②

ii) What type of bonding is present in this compound?

.... ①

TOTAL 9

2 Tim carries out an experiment to find out which of two metal oxides (A or B) is the better catalyst for the decomposition of hydrogen peroxide, H_2O_2.

10 cm^3 of hydrogen peroxide is measured into a conical flask and 40 cm^3 of water is added.

After 5 minutes no oxygen gas is collected.

0.5 g of A is added to the hydrogen peroxide in the flask. The volume of oxygen, O_2, produced is measured at intervals.

The results are shown in the table.

Time in minutes	0	1	2	4	6	8	10	12	14
Volume of gas collected in cm^3	0	21	32	45	52	57	59	60	60

a)   Write a balanced symbol equation for the decomposition of hydrogen peroxide.

.... ①

b)   Plot the results in the table on the grid. Draw the best line.

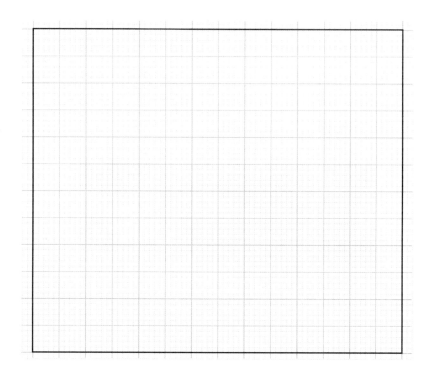

③

c)   Finish the detailed account of the experiment including how to process the results.

....

....

....

.... ④

d)   How could you tell from the graph which was the better catalyst?

.... ①

e) i) What is the maximum volume of oxygen obtained in the experiment with oxide A?

....cm³ ①

   ii) The concentration of a hydrogen peroxide solution is sometimes given in terms of its **volume strength**.

       A 10 volume solution produces 10 cm³ of oxygen when 1 cm³ of the solution is decomposed. What is the volume strength of the original hydrogen peroxide in this experiment?

.... ②

f)   Suggest reasons why:

   i) a 100 volume hydrogen peroxide decomposes faster than a 10 volume solution

.... ①

ii) an inhibitor is added to hydrogen peroxide after manufacture

.... ①

iii) hydrogen peroxide should be stored in dark glass bottles

.... ①

iv) hydrogen peroxide should **not** be stored in a glass bottle with a tight-fitting screw cap.

.... ①

TOTAL 16

❸ This question is about energy changes in the reaction of hydrogen and oxygen to form water.

$$2H_2 + O_2 \rightarrow 2H_2O$$

The table gives bond energies you will need.

Bond	Bond energy in kJ/mol
H–H	436
O=O	496
H–O	463

a) i) How much energy is needed to break the bonds in one mole of oxygen molecules?

....kJ ①

ii) How much energy is needed to break the bonds in two moles of hydrogen molecules, H_2?

....kJ ①

b) What is the total energy needed to break all the bonds in the reactant molecules?

....kJ ①

c) How much energy is released to the surroundings when all the bonds form in the product molecules, $2H_2O$?

....kJ ①

d) Calculate the overall energy change for this reaction.

....kJ ②

e) Complete the energy level diagram for the reaction by adding **reactants**, **products** and **overall energy change**.

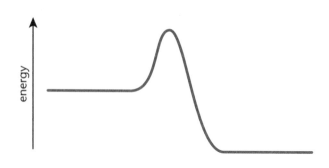

③

TOTAL 9

ANSWERS ON PAGE 40 ANSWERS ON PAGE 40 ANSWERS ON PAGE 40 ANSWERS ON PAGE 40

4 The table gives some information about halogen elements.

Halogen	Atomic number	Melting point in °C	Boiling point in °C	Atomic radius in nm	Electron arrangement
fluorine	9	−220	−188	0.071	2,7
chlorine	17	−101	−34	0.099	2,8,7
bromine	35	−7	58	0.114	2,8,18,7
iodine	53	114	183	0.133	2,8,18,18,7

a) i) How does the atomic radius of halogen elements change down the group?

.... ①

ii) Why is this?

.... ①

b i) How does the reactivity of halogen elements change down the group?

.... ①

ii) Explain this pattern.

....

.... ②

c) When chlorine gas is bubbled through a solution of potassium bromide, a reaction takes place.

i) What name is given to this type of reaction?

.... ①

ii) What would be seen when this reaction occurs?

.... ②

iii) Write a balanced equation for this reaction.

.... ③

d) Astatine (atomic number 85) is another halogen element. Predict some properties of astatine.

....

.... ②

TOTAL 13

5 The equation for the Haber process is shown below.

$$N_2 + 3H_2 \rightleftharpoons 2NH_3$$

The table gives the percentage of ammonia in equilibrium with nitrogen and hydrogen under different conditions.

Pressure in MN/m²	Percentage of ammonia at equilibrium Temperature in °C				
	300	400	500	600	700
0.101	2.18	0.44	0.13	0.05	0.02
1.01	14.7	3.85	1.21	0.49	0.23
3.04	31.8	10.7	3.62	1.43	0.66
10.1	51.2	25.1	10.4	4.52	2.18
20.3	62.8	36.3	17.6	8.2	4.10
101	92.6	79.8	57.5	31.4	12.9

a) What is the effect on the position of equilibrium of:

   i) increasing pressure

   .... ①

   ii) increasing temperature?

   .... ①

b) Why do manufacturers choose 450°C and 10.1 MN/m² rather than conditions that give better yields?

   ....

   ....

   .... ③

c) What effect does the iron catalyst have on the yield of ammonia?

   .... ①

TOTAL 6

6 Ali carried out some experiments to find the effect of the enzyme catalase on the decomposition of hydrogen peroxide.

$$2H_2O_2(aq) \rightarrow 2H_2O(l) + O_2(g)$$

Each time he used the same quantities of hydrogen peroxide and catalase in the apparatus shown in the diagram on the next page. He measured the volume of gas collected after 30 seconds. He carried out the experiments at different temperatures.

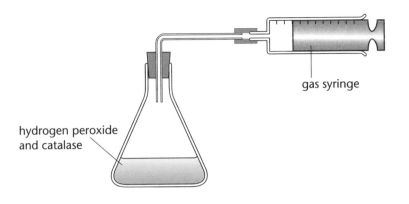

hydrogen peroxide
and catalase

gas syringe

The results are shown in the table.

Temperature of reactants in °C	Volume of gas collected in cm^3
0	4
10	7.5
20	14
30	30
35	40
40	44.5
50	23
60	0
70	0
80	0
90	0

a)   Describe and explain the results of his experiment.

....

....

....

....

....   ⑤

b)   Enzymes in yeast can ferment glucose.

i) What are the products of this process?

....   ②

ii) What conditions are needed for fermentation to take place?

....

....   ②

TOTAL 9

Patterns of behaviour *(vertical side text)*

1 a) Barium, strontium, calcium, magnesium ❶

b) i) Reacts very slowly with cold water ❶

ii) $Be + 2H_2O \rightarrow Be(OH)_2 + H_2$ ❸
(One mark for reactants, one mark for products and one mark for balancing.)

c) Under petroleum oil ❶

d) i) Barium chloride ❶
$BaCl_2$ ❶
ii) Ionic bonding ❶

2 a) $2H_2O_2 \rightarrow 2H_2O + O_2$ ❶
b) Correct scales and axes ❶
Correct plotting ❶
Best line ❶

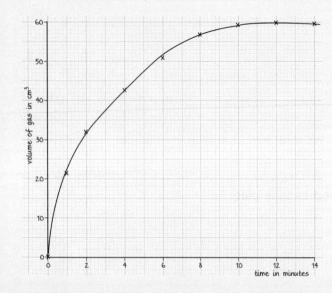

c) Fresh $10cm^3$ of hydrogen peroxide and $40cm^3$ of water.
Add 0.5 g of B.
Measure the volume of gas collected each minute.
Plot a graph on the same grid. ❹
d) The steeper graph is the faster reaction/the reaction that finishes first ❶
e) i) $60\,cm^3$ ❶
ii) $10\,cm^3$ hydrogen peroxide produces $60\,cm^3$ of oxygen ❶
6 volume ❶
f) i) The solution is more concentrated, more collisions between particles so faster reaction. ❶
ii) An inhibitor is a negative catalyst. It slows down the decomposition of hydrogen peroxide. ❶
iii) Light would speed up reaction. Dark glass does not let light through the bottle. ❶
iv) If hydrogen peroxide decomposes, pressure builds up inside the bottle. An explosion could result. ❶

3 a) i) 496 kJ ❶
ii) 436 x 2 = 872 kJ ❶
b) 496 + 872 = 1368 kJ ❶
c) 4 x 463 = 1852 kJ ❶
d) 1368 – 1852 ❶
= –484 kJ ❶
e) (One mark for each correct label.) ❸

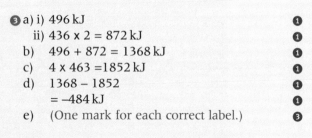
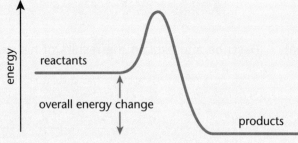

4 a) i) Increases down the group. ❶
ii) More electron shells ❶
b) i) Reactivity decreases down the group ❶
ii) Atom attracts electron ❶
Smaller atom has larger attracting power ❶

c) i) Displacement reaction (or redox reaction) **❶**
   ii) The colourless solution turns brown **❷**
   iii) $2KBr + Cl_2 \rightarrow 2KCl + Br_2$ **❸**

d) Astatine is a black solid. Less reactive than iodine. **❷**

❺ a) i) Moves to the right **❶**
   ii) Moves to left **❶**

b) Increasing pressure is extremely expensive. **❶**

Decreasing temperature increases yield but decreases rate. **❶**
A compromise has to be made. **❶**
c) No effect **❶**

❻ a) Any five points from the following: **❺**
   the rate of reaction increases up to about 40°C
   increasing temperature makes particles move faster producing more successful collisions
   there is a maximum value
   above 37°C the rate decreases
   enzymes denatured above 60°C
   no reaction occurs
b) i) Ethanol (alcohol) and carbon dioxide **❷**
   ii) Temperature about 37°C **❶**
   Absence of air **❶**

Water

To revise this topic more thoroughly, see Chapter 4 in *Letts Revise GCSE Chemistry Study Guide*.

 Try this sample GCSE question and then compare your answers with the Grade C and Grade A model answers on the next page.

On the label of a bottle of mineral water the composition of the water is given. The water contains temporary and permanent hardness. Here is part of the label.

Ions present	Concentration in g/litre
calcium Ca^{2+}	0.12
magnesium Mg^{2+}	0.02
sodium Na^+	0.01
potassium K^+	0.01
sulphate SO_4^{2-}	0.15
hydrogen carbonate HCO_3^-	0.26

a   Write down the name and formula of a compound in this water that causes permanent hardness.

Name...formula... [2]

b   Temporary hardness is caused by dissolved calcium hydrogencarbonate.

  (i) Explain how this compound gets into the water.

....

....

.... [3]

  (ii) Explain how the hardness is removed from this water by boiling.

....

....

.... [3]

c   Soap contains sodium stearate, $C_{17}H_{35}COO^-Na^+$.

Explain why using soap in hard water produces scum.

....

....

.... [3]

(Total 11 marks)

These two answers are at grades C and A. Compare which one your answer is closest to and think how you could have improved it.

GRADE C ANSWER

Holly

a   Calcium sulphate ✓   CaSO₄ ✓
b (i) Calcium carbonate in rocks ✓
       is washed out by rain ✓ ✗
(ii) Calcium hydrogencarbonate
      decomposes when heated and the
      products are lost as gases ✓ ✗ ✗
c   Sodium stearate reacts with
      hard water to form scum ✓ ✗ ✗

Holly loses marks here because there is a lack of detail.

Again Holly loses marks because there is a lack of detail. She might be able to write a simple word equation, e.g. Sodium stearate+ calcium ions → calcium stearate+ sodium ions. This would have scored at least one extra mark.

6 marks = Grade C Answer

Grade booster ···⟩ move a C to a B

Holly lacks detail in some of her answers especially in questions worth 3 marks. She should make sure she makes at least 3 points in a question worth 3 marks.

GRADE A ANSWER

Damini

a   Calcium or magnesium sulphates ✓ CaSO4 ✓
b (i) Calcium carbonate reacts with water and
       carbon dioxide to produce calcium
       hydrogencarbonate. This dissolves in water. ✓✓✓
(ii) Calcium hydrogen carbonate decomposes on
      boiling producing insoluble scale. ✓✓ ✗
c   Sodium stearate reacts with calcium in water
      to form calcium stearate. This is insoluble in
      water and forms scum.
      $2C_{17}H_{35}COO^-Na^+ + Ca^{2+} \rightarrow (C_{17}H_{35}COO^-)_2Ca$ ✓✓✓

Damini has lost a mark here as her answer lacks detail. Ideally here she could write an equation

$Ca(HCO_3)_2 \rightarrow CaCO_3 + H_2O + CO_2$

10 marks = Grade A Answer

Grade booster ···⟩ move A to A*

This excellent answer still has opportunities to include relevant chemical equations in b (i) and (ii). An A* candidate should include equations whenever relevant.

43

Water

1 There are two brands of bottled water on sale in Egypt, SIWA and BARAKA.
The table gives the concentrations of different ions dissolved in each.

Ion	Concentration in mg/dm^3	
	SIWA	BARAKA
calcium	8	60
magnesium	9	26
sodium	42	62
potassium	18	5
hydrogencarbonate	122	325
sulphate	18	50
chloride	36	50
silica	10	20
fluoride	0.5	nil

a) Which brand contains more hardness? Explain your answer.

....

.... ①

b) One water is taken from deep underground limestone rocks.
Which sample is this? Explain your answer.

....

.... ①

c) i) What is the benefit of having fluoride ions in water?

.... ①

ii) Why do some people object to it being added to water?

....

.... ①

d) When a sample of BARAKA is boiled, the water goes slightly cloudy.
Explain why this happens and write a balanced symbol equation for the reaction.

....

....

.... ④

e) Both samples are heated with ozone, a reactive form of oxygen, before being bottled.

   i) Suggest why the water is treated with ozone.

   .... ①

   ii) What is used to treat tap water to do the same job?

   .... ①

2 The table contains the results of experiments carried out adding soap solution to different samples of water until a lasting lather is formed. The water samples were distilled water and samples labelled W, X, Y and Z. In each case 100 cm³ samples of water were used and the soap was added in small measured portions.

Sample	Reading before/cm³	Reading after/cm³	Volume of soap added/cm³
distilled water	0.0	0.5	0.5
W	0.5	7.0	6.5
X	7.0	18.5	
Y	18.5	30.0	
Z	30.0	30.5	
W after boiling	30.5	34.0	
X after boiling	34.0	45.5	
Y after boiling	45.5	46.0	
Z after boiling	46.0	46.5	

a) Write down two things that ensured a fair test.

   i) ....

   ii) .... ②

b) Write down the name of a piece of apparatus suitable for:

   i) measuring out 100 cm³ samples of water.. ①

   ii) adding soap solution in small measured portions.. ①

c) Complete the table. ①

d) Which sample of water contained only temporary hardness? Explain your answer.

   ....

   .... ②

e) Which sample of water contained only permanent hardness? Explain your answer.

   ....

   .... ②

f) Which sample of water contained temporary and permanent hardness? Explain your answer.

   ....

   .... ②

3 Jenny carried out an experiment to find the solubility of sodium chloride in water at room temperature. Here are her results:

Mass of evaporating basin = 54.65 g

Mass of evaporating basin + sodium chloride solution = 122.25 g

Mass of evaporating basin + solid sodium chloride = 72.45 g

a) Calculate the solubility of sodium chloride from Jenny's results.

④

Solubility = ... g per 100 g of water

b) Within the whole class the results of the solubility of sodium chloride in water varied from 28 g/100 g of water to 40 g/100 g water.

The result in the data book was 36 g/100 g of water.

Assume that all weighings and temperature measurements were correctly and accurately made.

i) Suggest why a student might get a result less than 40 g/100 g of water.

.... ①

ii) Suggest why a student might get a value more than 40 g/100 g of water.

.... ①

iii) Suggest an improvement to the experiment that might give better results.

....

.... ①

TOTAL 7

4 The table gives the solubilities (in g/100 g of water) of three compounds in water at different temperatures.

Compound	Temperature (°C)						
	0	20	30	40	60	80	100
Barium chloride $BaCl_2.2H_2O$	32	36	38	41	46	52	59
Ammonium chloride NH_4Cl	30	37	41	46	55	66	77
Sodium carbonate crystals $Na_2CO_3.10H_2O$	7	21	40	48	47	46	45

a) Plot these three sets of data on the grid. Plot temperature on the x-axis. Draw three best-fit lines and label them with the names of the compounds.

④

b) Which substance is most soluble at 25°C?

.... ①

c) At which temperature is the solubility of barium chloride 50 g/100 g of water?

.... ①

A saturated solution of ammonium chloride at 80°C contains 100 g of water.

d) The solution was allowed to cool to 20°C. What mass of ammonium chloride would crystallise out? Show your working.

②

Mass .... g

e) What is unusual about the solubility of sodium carbonate crystals? Suggest an explanation for this.

....

....

.... ②

TOTAL 10

5 The table gives the solubility of some common gases in water (in g/100 g of water).

Gas	Temperature /°C				
	0	20	40	60	80
ammonia NH_3	89.5	53.1	30.7		
carbon dioxide CO_2	0.335	0.169	0.097	0.058	
hydrogen H_2	1.92×10^{-4}	1.60×10^{-4}	1.38×10^{-4}	1.18×10^{-4}	0.79×10^{-4}
sulphur dioxide SO_2	0.228	0.113	0.054		
hydrogen chloride HCl	82.3	72.1	63.3	56.1	
nitrogen N_2	2.94×10^{-3}	1.90×10^{-3}	1.39×10^{-3}	1.05×10^{-3}	
oxygen O_2	6.9×10^{-3}	4.34×10^{-3}	3.08×10^{-3}	2.27×10^{-3}	1.38×10^{-3}

a) Which two gases are very soluble in water?

.... ②

b) How does the solubility of gases in water change with temperature at constant pressure?

.... ①

c) Fish live in water using dissolved oxygen. Pollution of water reduces dissolved oxygen.

Why are fish soon affected by pollution?

.... ①

d) A saturated solution of carbon dioxide at 0°C is heated to 60°C.

What mass of carbon dioxide would be expelled from 100 g of solution?

Show your working.

②

Mass .... g

TOTAL 6

6 Tap water is taken from underground sources, clean rivers and reservoirs.

Write an account of the stages for treating this water before it can be used as tap water. Explain why each stage is necessary.

....

....

....

....

....

....

....

....

.... ⑥＋①

TOTAL 7

7 You are given two liquids. One is pure water and the other is sodium chloride solution.

a) Explain why anhydrous copper(II) sulphate cannot be used to tell which of these liquids is which.

....

.... ②

b) Suggest three ways of deciding which liquid is which. You must not taste the liquids.

 i) ....

....

 ii) ....

....

 iii) ....

.... ⑥

TOTAL 8

8 There are two types of detergent – soaps and soapless detergents.

a) What is the advantage of a soapless detergent in a hard water area?

.... ①

b) How are soap and soapless detergent molecules similar in structure?

....

.... ②

c) Describe how a soap or soapless detergent removes dirt from cloth.

....

....

....

....

.... ④

TOTAL 7

① a) Baraka (no mark for just the answer)
High concentration of calcium and
magnesium ions. ❶

b) Baraka (no mark for just the answer)
Limestone (calcium carbonate) reacts with
water and carbon dioxide to produce soluble
calcium hydrogencarbonate. ❶

c) i) Reduces dental decay ❶
ii) Fluoride is poisonous in large quantities/
ignorance of science ❶

EXAMINER'S TIP

*This is an Ideas and Evidence question. There
is a range of acceptable answers.*

d) Temporary hardness (calcium
hydrogencarbonate) decomposes ❶
Calcium carbonate does not dissolve in
water so it goes cloudy ❶
$Ca(HCO_3)_2 \rightarrow CaCO_3 + H_2O + CO_2$ ❷

EXAMINER'S TIP

*One mark for reactants and one mark for
products. No mark for balancing.*

e) i) Kill bacteria ❶
ii) Chlorine ❶

② a) Equal volume of water samples ❶
Same soap solution ❶
b) i) Measuring cylinder ❶
ii) Burette ❶
c) 11.5, 11.5, 0.5, 3.5, 11.5, 0.5, 0.5
(All required) ❶
d) Y ❶
All of hardness removed (11.5 cm³
before, 0.5 cm³ after) ❶
e) X ❶
Same volume of soap before and after
boiling ❶
f) W ❶
Some of hardness removed but not all ❶

EXAMINER'S TIP

*This question requires little knowledge but a
great deal of understanding. Make sure you
use the data fully.*

③ a) Mass of sodium chloride = 17.8 g ❶
Mass of water = 49.8 g ❶
Solubility = $\dfrac{17.8 \times 100}{49.8}$ ❶
= 35.7 g ❶

b) i) The saturated solution made to evaporate
was not in fact saturated. ❶
ii) Not all the water removed after
evaporation. ❶
iii) After weighing the sodium chloride,
heat the evaporating basin again and then
re-weigh. If all the water is removed the
mass should be unchanged. ❶

EXAMINER'S TIP

*This question is testing your evaluation skills.
You should have practice doing this in
Coursework. There are other possibilities, e.g.
ways of ensuring that the solution is saturated
or avoiding transferring any solid sodium
chloride to the evaporating basin before
evaporation.*

④ a)

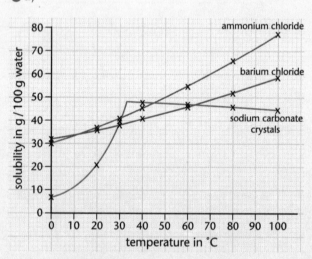

Correct scales and axes ❶
Plotting ❷
Three graphs ❶
b) Ammonium chloride ❶
c) 68–72°C ❶

d) 66–37 ❶
29 g ❶

e) The curve is not a smooth curve but two curves that meet at a point. ❶
Sodium carbonate can exist in two forms. ❶

❺ a) Hydrogen chloride and ammonia ❷
b) It decreases with increasing temperature ❶

c) There is very little dissolved oxygen so any reduction is serious ❶
d) 0.335–0.058 g ❶
= 0.277 g ❶
[N.B. Correct to 0.28 g]

❻ Any three of the following:
(one mark for method and one for explanation)
1 Filtering or screening – to remove solids
2 Adding lime – remove excess acidity
2 Adding aluminium ions (aluminium sulphate) – to coagulate colloids
2 Adding carbon (charcoal) – to remove odours
3 Adding chlorine – kill bacteria
4 Adding sulphur dioxide to remove excess chlorine.
(In addition one mark for Quality of Written Communication. To get this you must get three methods in the correct order. 1, 2, 3 or 2 ,3, 4 or 1, 3, 4 would be suitable.) ❻+❶

❼ a) Anhydrous copper(II) sulphate turns white to blue with both. ❶
It is a test for water present not just for pure water. ❶

b) i) Evaporate to dryness ❶
Pure water leaves no residue, sodium chloride solution leaves solid sodium chloride. ❶
ii) Conductivity of electricity ❶
Pure water does not conduct electricity, sodium chloride solution does. ❶
iii) Test boiling point ❶
Pure water boils at 100ºC, sodium chloride solution boils at higher temperature, e.g. 105ºC ❶

❽ a) Soaps form scum in hard water ❶
b) Both have ionic head ❶
and hydrophobic tail ❶
c) Tails of detergent molecules stick into grease ❶
Water loving heads of detergent molecules attracted by water molecules ❶
Agitation lifts off grease ❶
Grease kept suspended in water ❶

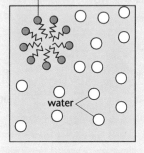

cluster of detergent molecules

water

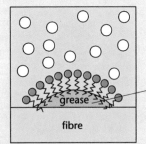

tails of detergent stick in grease

grease

fibre

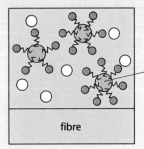

grease suspended in water

fibre

Acids, bases and salts

To revise this topic more thoroughly, see Chapter 5 in *Letts Revise GCSE Chemistry Study Guide*.

 Try this sample GCSE question and then compare your answers with the Grade C and Grade A model answers on the next page.

Sulphuric acid, H_2SO_4, is a dibasic acid.

a What is meant by dibasic acid?

.... [1]

b Describe and explain simple tests to show that dilute sulphuric acid

  (i) has a pH value of 1

.... [1]

  (ii) contains sulphate ions.

....

.... [2]

c Describe two chemical properties of dilute sulphuric acid that are typical of an acid. In each case include a balanced equation.

  (i) ....

....

....

  (ii) ....

....

.... [10]

d Write an ionic equation for one of the reactions in **c**.

.... [3]

(Total 17 marks)

GRADE C ANSWER

Ashish

a   Acid with two replaceable
     hydrogens ✓

b (i) Add universal indicator ✗

(ii) Add dilute sulphuric acid and
      barium chloride soln. ✗
      White precipitate ✓

c (i) Add magnesium metal ✓
       Magnesium dissolves ✗
       $Mg + H_2SO_4 \rightarrow MgSO_4 + 2H$
                              ✓ ✗ ✓

(ii) Add copper oxide ✓
      $CuO + H_2SO_4 \rightarrow CuSO_4 + H_2O$
                              ✓ ✓ ✓

d $Mg + SO_4^{2-} + 2H^+ \rightarrow Mg^{2+} + SO_4^{2-} + H_2$
                              ✓ ✓ ✗

The one mark available is for both the name of the indicator and the colour it changes too.

Ashish has made the mistake that many make of using sulphuric acid instead of hydrochloric acid to acidify. Doing this will always give a positive sulphate test.

This is not the simplest equation because it has sulphate ions on both sides of the equation.

Magnesium does not dissolve, it reacts. However, the important observation is the formation of bubbles of hydrogen gas.

Hydrogen exists as molecules H_2.

The examiner has not penalised Ashish for writing copper oxide instead of copper(II) oxide.

11 marks = Grade C Answer

Grade booster ⋯⟩ move a C to a B

There is a lack of detail in places in this answer and a couple of common errors. In chemical reactions, such as c, include observations. The examiner wants to know that you have seen these reactions.

GRADE A ANSWER

Hazel

a   Two hydrogens in each molecule can be
     replaced ✓

b (i) Add universal indicator, turns red ✓

(ii) Add dilute hydrochloric acid and barium
      chloride soln. ✓
      White precipitate ✓

c (i) Add copper(II) oxide ✓ Forms blue solution
       of copper(II) sulphate ✓
       $CuO + H_2SO_4 \rightarrow CuSO_4 + H_2O$ ✓ ✓ ✓

(ii) Add sodium hydrogencarbonate ✓ Fizzes ✓
      $NaHCO_3 + H_2SO_4 \rightarrow Na_2SO_4 + H_2O + CO_2$
                              ✓ ✓ ✗

d $O^{2-} + 2H^+ \rightarrow H_2O$ ✓ ✓ ✓

Here the error is in balancing the equation.

16 marks = Grade A Answer

Grade booster ⋯⟩ move A to A*

This is an excellent answer giving appropriate details throughout. The only error is in balancing one equation in c. This is easy to do under examination conditions. Check every equation before moving on.

Acids, bases and salts

53

Acids, bases and salts

1 Sam prepares crystals of sodium sulphate starting with sodium hydroxide solution and dilute sulphuric acid.

a) Write a balanced symbol equation for the reaction.

.... ③

b) Describe the method of preparation Sam should use.

....

....

....

....

....

....

.... ⑤+①

TOTAL 9

2 Sodium hydrogensulphate, $NaHSO_4$, is a white crystalline solid. It is used in Chemistry sets for children in place of dilute sulphuric acid. In solution it reacts in a similar way to sulphuric acid.

a) Why is it better for manufacturers to use sodium hydrogensulphate than dilute sulphuric acid?

.... ①

b) Write balanced symbol equations for the reaction of sodium hydrogensulphate with:

i) magnesium ribbon

.... ③

ii) copper(II) oxide

.... ③

iii) sodium hydrogencarbonate.

.... ③

c) Write an ionic equation for one of the reactions in b).

.... ③

TOTAL 13

3 The table contains information about the reactions of dry citric acid crystals and an aqueous solution of citric acid.

Test	Dry citric acid crystals	Aqueous solution of citric acid
dry universal indicator paper	green, pH 7, neutral	red–pH less than 4–acidic
add magnesium ribbon	no bubbles of hydrogen	bubbles of hydrogen gas
add sodium carbonate crystals	no bubbles of carbon dioxide gas	bubbles of carbon dioxide gas

a) What can be concluded from these results?

.... ①

b) Citric acid has the formula

$$CH_2COOH$$
$$|$$
$$C(OH)COOH$$
$$|$$
$$CH_2COOH$$

i) What is the molecular formula of citric acid?

.... ①

ii) Suggest what changes take place when citric acid is added to water.

.... ①

c) Sulphuric acid is a strong acid and citric acid is a weak acid. Explain the terms strong and weak with reference to sulphuric acid and citric acid.

....

....

.... ②

TOTAL 5

---

4 Lead(II) sulphate is an insoluble salt.

a) Write down the name of a method used to prepare an insoluble salt.

.... ①

b) Describe how a pure sample of lead(II) sulphate could be prepared.

....

....

....

.... ④+①

c) Write a balanced symbol equation for the formation of lead(II) sulphate.

.... ③

d) Lead(II) sulphate used to be used as a white pigment in household paints.
Suggest why it is no longer used.

.... ①

TOTAL 10

5 Three bottles contain different colourless solutions. The bottles are labelled X, Y and Z.

One bottle contains dilute hydrochloric acid, one contains sulphuric acid and one contains sodium sulphate.

a) Describe chemical tests you could use to tell which is which.

....
....
....
....
....
.... ⑥

b) Write an ionic equation for one of the reactions in a).

.... ③

TOTAL 9

6

a) Describe how a sample of zinc sulphate crystals could be prepared from zinc carbonate.

....
....
....
....
....
....
....
.... ⑥+①

b) Write a balanced symbol equation for the reaction in a).

.... ③

TOTAL 10

7 Ethanedioic acid, $C_2O_4H_2$ is a weak, dibasic acid.

a) Write an equation showing what happens when solid ethanedioic acid is dissolved in water.

.... ③

b) Write down the name of the scientist who introduced the ideas of dissociation.

.... ①

c) Write down the formula of the salt produced when sodium hydroxide reacts with ethanedioic acid.

.... ①

TOTAL 5

8

a) Finish the following general equations.

acid + alkali → .... + ...

acid + carbonate → .... + ... + ... ②

b) Give two examples of neutralisation being used in industry.

i) ....

ii) .... ②

TOTAL 4

1 a) $2NaOH + H_2SO_4 \rightarrow Na_2SO_4 + 2H_2O$ **❸**
(One mark for correct formulae of reactants, one for products and one for balancing.)

b) Any five points from the following:
measure out 25.0 cm³ of sodium hydroxide with a pipette
add indicator
add sulphuric acid from a burette in small portions until neutral
record volume of acid added
repeat without indicator
evaporate until a small volume of solution remains
leave to cool
(In addition one mark is available for Quality of Written Communication. For this your answer has to be written so the events are in a correct and logical order.) **❺+❶**

EXAMINER'S TIP

This question is about understanding a practical process. Candidates have difficulty in giving enough detail in b). They might mix acid and alkali but not attempt to get the correct reacting amounts. This could score the last three marks only.

2 a) Solid is easier to pack and transport/does not spill. **❶**

b) i) $2NaHSO_4 + Mg \rightarrow MgSO_4 + Na_2SO_4 + H_2$ **❸**
ii) $2NaHSO_4 + CuO \rightarrow CuSO_4 + Na_2SO_4 + H_2O$ **❸**
iii) $NaHSO_4 + NaHCO_3 \rightarrow Na_2SO_4 + H_2O + CO_2$ **❸**

c) One from:
$Mg + 2H^+ \rightarrow Mg^{2+} + H_2$
$O^{2-} + 2H^+ \rightarrow H_2O$
$HCO_3^- + H^+ \rightarrow H_2O + CO_2$ **❸**

EXAMINER'S TIP

Each time in b) and in c) there is one mark for correct formulae of reactants, one for products and one for balancing.

3 a) A solid such as citric acid only shows acidic properties when water is present. **❶**

b) i) $C_6H_8O_7$ **❶**
ii) Three hydrogen atoms are lost as hydrogen ions **❶**

EXAMINER'S TIP

The equation below is useful but not essential. It shows citric acid is a weak acid.

$$C_6H_8O_6 \rightleftharpoons C_6H_5O_6{}^{3-} + 3H^+$$

c) Sulphuric acid is completely ionised in aqueous solution. **❶**
$[H_2SO_4 \rightarrow 2H^+ + SO_4{}^{2-}]$
Citric acid is only partially ionised. **❶**

4 a) Precipitation **❶**
b) Any four points from the following:
mix together suitable solutions of lead(II) ions, e.g lead(II) nitrate
and a suitable solution of a sulphate, e.g sulphuric acid or sodium sulphate
white precipitate formed
filter
wash and dry
(In addition one mark is available for Quality of Written Communication. For this your answer has to be written so the events are in a correct and logical order.) **❹+❶**

c) $Pb(NO_3)_2 + H_2SO_4 \rightarrow PbSO_4 + 2HNO_3$ **❸**
d) Lead is toxic and can cause problems if children chew painted items or when sanding down old paintwork. **❶**

5 a) One mark for each point:
add barium chloride solution to each solution
sodium sulphate and sulphuric acid produce a white precipitate
the one that does not produce a precipitate is hydrochloric acid
add sodium hydrogencarbonate to the two remaining solutions
sulphuric acid fizzes and produces a colourless gas (carbon dioxide)
sodium sulphate solution produces no precipitate **❻**

EXAMINER'S TIP

There are other methods that could be used here.

b) $Ba^{2+} + SO_4{}^{2-} \rightarrow BaSO_4$
or $H^+ + HCO_3^- \rightarrow H_2O + CO_2$ **❸**

Acids, bases and salts

6 a) Add sulphuric acid to beaker ❶
Add solid zinc carbonate in small portions,
   stirring after each addition ❶
until some solid remains unreacted ❶
filter off excess ❶
evaporate until a small volume of
   solution remains ❶
leave to cool and zinc sulphate crystals
   form ❶
(In addition one mark is available for
Quality of Written Communication. For
this your answer has to be written so the
events are in a correct and logical
order.) ❶

b) $ZnCO_3 + H_2SO_4 \rightarrow ZnSO_4 + H_2O + CO_2$ ❸

7 a) $C_2O_4H_2 \rightleftharpoons C_2O_4^{2-} + 2H^+$ ❸

EXAMINER'S TIP

*Here one mark is for reactants and products,
one for balancing and one for the reversible
reaction sign.*

b) Arrhenius ❶

EXAMINER'S TIP

*The work of Arrhenius is mentioned
particularly in AQA specifications.*

c) $Na_2C_2O_4$ ❶

8 a) acid + alkali → salt + water ❶
acid + carbonate → salt + water + carbon
dioxide ❶

b) Neutralising acidity in soil to increase crop
yields ❶
Removing sulphur dioxide from chimneys
of coal-fired power stations ❶

Acids, bases and salts

Metals and redox reactions

To revise this topic more thoroughly, see Chapter 6 in *Letts Revise GCSE Chemistry Study Guide*.

Try this sample GCSE question and then compare your answers with the Grade C and Grade A model answers on the next page.

Lead is extracted from lead(II) sulphide.

a   Lead(II) sulphide is roasted in air to produce lead(II) oxide and sulphur dioxide gas. The lead(II) oxide is then heated in a blast furnace with coke (carbon).

(i) Write a symbol equation for the reaction of lead(II) sulphide and oxygen.

   $2PbS + 3O_2 \longrightarrow 2PbO + 2SO_2$ **[3]**

(ii) Explain why allowing sulphur dioxide to escape from this process into the atmosphere causes environmental damage.

   SO_2 causes acid rain. sulphuric acid is produced when SO_2 is dissolved in water.   **[3]**

(iii) Suggest why lead is often extracted in a factory close to another factory producing sulphuric acid.

   SO_2 is used in the manufacture of sulphuric acid.   **[1]**

b   The equation for the reaction that takes place when lead(II) oxide is heated with carbon in the blast furnace is given below.

$$2PbO + C \rightarrow Pb + CO_2$$

What is the function of coke in the process?

   acts as a reducing agent.   **[1]**

c   The lead produced in this process is mixed with small amounts of silver.

(i) What name is given to a mixture of metals?

   Alloy   **[1]**

(ii) Suggest why is it worthwhile extracting silver from the lead.

   The profit of recovering the silver compensates for the cost of extraction   **[1]**

(Total 10 marks)

GRADE C ANSWER

Mohammed has given the wrong formula for lead(II) sulphide. However the products are correct and the balancing of the wrong equation is worth a mark.

Mohammed

a (i) $2PbS_2 + 5O_2 \rightarrow 2PbO + 4SO_2$ ✗✓✓

(ii) Causes acid rain ✗✗✓

(iii) Sulphur dioxide is used to make sulphuric acid. ✓

b   Oxidising agent ✗

c (i) Alloy ✓

(ii) The recovered silver is worth far more than the costs of removing it from the lead. ✓

The answer given here cannot score more than one mark and there are three marks available.

6 marks = Grade C Answer

Grade booster ····▷ move a C to a B

Mohammed needed to take more notice of the marks available in **a (ii)**. If a question is worth 3 marks, Mohammed must make at least three points.

GRADE A ANSWER

In this answer there should have been reference to the job of oxygen here. Water dissolves sulphur dioxide to form sulphurous acid and this is oxidised by oxygen in the air into sulphuric acid.

Usha

a (i) $2PbS + 3O_2 \rightarrow 2PbO + 2SO_2$ ✓✓✓

(ii) Sulphur dioxide dissolves in rain water to produce sulphuric acid. ✓✗ Causes acid rain ✓

(iii) Sulphur dioxide is used to make sulphuric acid. ✓

b   Reducing agent. ✓

c (i) Alloy ✓

(ii) Silver is unreactive - low in the reactivity series. ✗

This answer is incorrect. It needs an economic answer. The statement is true but does not explain why it is worthwhile to do this.

8 marks = Grade A Answer

Grade booster ····▷ move A to A*

This is a very good answer but misses out in just a couple of places. In **c (ii)** the answer given does not match the question. The word *worthwhile* is pointing Usha to an economic answer.

Metals and redox reactions

QUESTION BANK

1 Explain the meaning of the terms oxidation and reduction in terms of electron transfer.

a) *oxidation is loss of electrons.*
Reduction is gain of electrons. (2)

b) Explain whether oxidation or reduction has taken place in each of the following reactions. In one case no oxidation or reduction has taken place. (5)

  i) $Cl_2(g) + 2e^- \rightarrow 2Cl^-(aq)$
  Reduction

  ii) $CO_3^{2-}(aq) + 2H^+(aq) \rightarrow H_2O(l) + CO_2(g)$
  no oxidation or reduction

  iii) $Ca(s) \rightarrow Ca^{2+}(aq) + 2e^-$
  Oxidation

  iv) $H_2O_2(aq) \rightarrow 2H^+(aq) + O_2(g) + 2e^-$
  oxidation

  v) $MnO_4^-(aq) + 8H^+(aq) + 5e^- \rightarrow Mn^{2+}(aq) + 4H_2O(l)$
  reduction

c) Write an ionic equation for the reaction of acidified potassium manganate(VII) and hydrogen peroxide. Include state symbols. Use the equations in b) to help you.

  $5H_2O_2 + 2MnO_4^- + 6H^+ \rightarrow 2Mn^{++} + 8H_2O + 5O_2$ (4)

TOTAL 11

2 Chromium can be extracted from chromium(III) oxide by heating a powdered mixture of chromium(III) oxide with magnesium.

a) What is the function of the magnesium in this reaction?
reducing agent (1)

b) How must the position of chromium compare with the position of magnesium in the reactivity series?
chromium must be lower in the reactivity series (1)

c) Write a balanced symbol equation for the reaction of chromium(III) oxide with magnesium.
$Cr_2O_3 + 3Mg \rightarrow 2Cr + 3MgO$ (3)

TOTAL 5

3 Steel is an alloy.

a) Explain what is meant by the term alloy.
a mixture of metal elements (1)

b) Molten iron from a blast furnace contains 4% carbon.

Describe the processes involved in turning this into steel.

....

.... ②

c) A lot of the steel produced is in the form of stainless steel.

i) What is added to steel to make stainless steel?

................*chromium*... ①

ii) What advantage does stainless steel have over other forms of steel?

............*does not corrode*.. ①

d) Write down the name of another alloy and give the main constituents of this alloy.

................*Brass*.. ②

TOTAL 7

④ An experiment was carried out with rods of copper, silver, zinc, iron and magnesium.

The apparatus is set up as shown in the diagram.

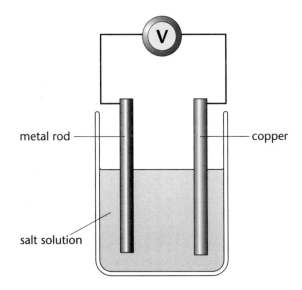

a) What would be the voltage on the voltmeter if the metal rod shown on the diagram was copper?

.... ①

b) Which metal would produce a negative reading on the voltmeter?

.... ①

c) Which metal would produce the largest reading on the voltmeter?

.... ①

d) Suggest two reasons why this type of cell cannot be used in a torch.

i) ....

ii) .... ②

TOTAL 5

a)   What conditions are needed for nails to rust?

    ....   ②

Imran was investigating the rusting of iron nails.

He put nails into dishes of clear jelly containing a rusting indicator.

The more the nails rusted, the more the indicator turned blue.

jelly containing rusting indicator (no blue colour at start)

iron nail

Here are some of Imran's results after five days.

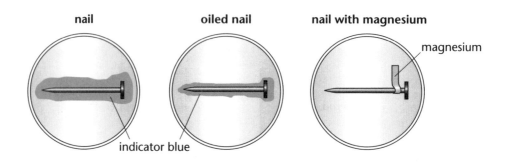

nail   oiled nail   nail with magnesium

                                           magnesium

indicator blue

b)   What **two** things can you conclude from this investigation?

 i)   ....

 ii) ....   ②

The order of some metals is shown in the table.

Reactivity series	
Most reactive	magnesium
	zinc
	iron
	tin
	copper
Least reactive	silver

Metals and redox reactions

c) Look at the results from the two dishes shown below.

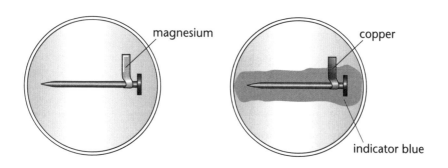

magnesium

copper

indicator blue

i) What is the relationship between the reactivity of the attached metal and the rusting of the iron nail?

....

.... ②

ii) Which one of the following dishes, W to Z, would show the results for a nail with a piece of tin attached?

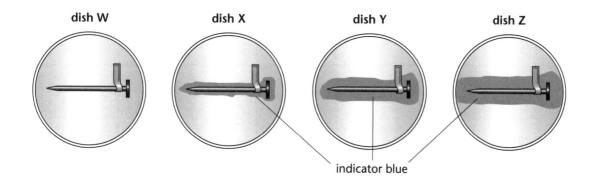

dish W   dish X   dish Y   dish Z

indicator blue

.... ①

d) Magnesium is an expensive metal.

Why is it attached to the iron legs of a pier which stand in sea water?

....

.... ②

TOTAL 9

Metals and redox reactions

6 A piece of zinc was added to copper(II) sulphate solution.

a) Describe the changes that you would observe.

....

.... ②

b) Write an ionic equation for the reaction taking place.

.... ③

c) Explain in terms of electron transfer why this is a redox reaction.

....

.... ②

TOTAL 7

ANSWERS ON PAGE 67 ANSWERS ON PAGE 67 ANSWERS ON PAGE 67 ANSWERS ON PAGE 67

1 a) Oxidation – loss of electrons ❶
Reduction – gain of electrons ❶
b) i) Reduction – gain in electrons ❶
ii) No oxidation or reduction ❶
iii) Oxidation – loss of electrons ❶
iv) Oxidation – loss of electrons ❶
v) Reduction – gain in electrons ❶
c) $5H_2O_2(aq) + 2MnO_4^-(aq) + 6H^+(aq) \rightarrow$
$2Mn^{2+}(aq) + 8H_2O(l) + 5O_2(g)$ ❹

EXAMINER'S TIP

In part c) one mark is for reactants, one for products, one for balancing and one for state symbols. This question is as difficult as any that can be set at GCSE level.

2 a) Reducing agent/to reduce chromium(III) oxide to chromium ❶
b) Magnesium must be higher in the reactivity series ❶
c) $Cr_2O_3 + 3Mg \rightarrow 3MgO + 2Cr$ ❸

3 a) Mixture of metals (or metal and carbon) ❶

EXAMINER'S TIP

In most cases the answer to this question is mixture of metals. However, the reference to steel means you should include the presence of carbon in steel.

b) Blow air or oxygen through molten iron to oxidise impurities. ❶
Add required amount of carbon to make steel. ❶
c) i) Add chromium or nickel ❶
ii) Does not rust ❶
d) Brass ❶
Copper and zinc ❶

EXAMINER'S TIP

There are other alloys you could give, e.g. solder, bronze.

4 a) Zero ❶
b) Silver ❶
c) Magnesium ❶

EXAMINER'S TIP

The biggest voltage is when the two metals have the biggest difference in reactivity.

d) Two from:
Very small voltages
Liquids would escape
Only produces a current when a very low resistance ❷

5 a) Both water and air/oxygen present ❷
b) i) Oiling the nail reduces rusting ❶
ii) No rusting when nail is in contact with magnesium ❶
c) i) Copper is less reactive than iron (lower than iron in the reactivity series) ❶
Increases the amount of rusting ❶
ii) X ❶

EXAMINER'S TIP

Tin is below iron so will speed up the rusting but not as much as with copper.

d) Sacrificial protection/to stop the legs rusting ❶
It would be expensive to replace iron legs ❶

6 a) Blue solution turns colourless ❶
Pink-brown solid formed ❶
b) $Zn + Cu^{2+} \rightarrow Cu + Zn^{2+}$ ❸
c) Zinc atom loses two electrons – oxidation ❶
Copper ion gains two electrons – reduction ❶

Metals and redox reactions

CHAPTER 7

Further carbon chemistry

To revise this topic more thoroughly, see Chapter 7 in *Letts Revise GCSE Chemistry Study Guide*.

 Try this sample GCSE question and then compare your answers with the Grade C and Grade A model answers on the next page.

a The simplest alcohol is methanol, CH_3OH.

Draw a 'dot and cross' diagram for a molecule of methanol.

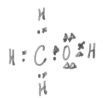

[3]

b Ethanol can be made from sugar. State the other reactant(s) required, the conditions necessary for a reaction and the main by-product. [4]

Reactant(s) ....

Conditions ....

Main by-product ....

c Write down two effects of ethanol on the human body. [2]

(i) ....

(ii) ....

d In addition to alcoholic drinks, give **two** uses of ethanol. [2]

(i) ....

(ii) ....

(Total 11 marks)

GRADE C ANSWER

Rana

a

H
×•
H ⦂ C ⦂ OH
×•
H
✓××

Rana has not done a 'dot and cross' diagram for the whole molecule.

b Enzymes ×
Room temperature × Keep oxygen out ✓
Carbon dioxide ✓
c i) Slows reaction times ✓
ii) Causes liver damage ✓
d i) Esters ×
ii) Solvent ✓

Enzymes is too vague. The answer requires the enzymes in yeast. Zymase would be correct. There is no mention of this process carried out in solution. The process is very slow at room temperature. The examiner is looking for 35–40°C.

Ethanol is not an ester. It is used to make esters. It is used as a solvent.

6 marks = Grade C Answer

Grade booster ⋯⟫ move a C to a B
Rana's answer lacks detail especially in the fermentation process. You cannot include too much detail.

GRADE A ANSWER

Prajawl

a

H
×• ○○
H ⦂ C ⦂ O ⦂ H
×• ○○
H
✓✓✓

The 'dot and cross' diagram is very good. Prajawl has identified three types of electrons – electrons from carbon atom (black circles) – electrons from hydrogen atom (crosses) – electrons from oxygen atom (white circles). This is more detail than expected.

b Glucose solution, yeast ✓
37°C ✓ absence of air ✓
carbon dioxide ✓
c i) Causes high blood pressure ✓
ii) Impairs judgement ✓
d i) Making esters ✓
ii) Paints ×

Ethanol is not used as a paint but to make paints.

10 marks = Grade A Answer

Grade booster ⋯⟫ move A to A*
The diagram here is excellent. It shows that in each covalent bond there is a pair of electrons – one electron from each atom.

Further carbon chemistry

Further carbon chemistry

1 Ethanol was dehydrated by passing the vapour over strongly heated broken china. The diagram shows the apparatus that was used.

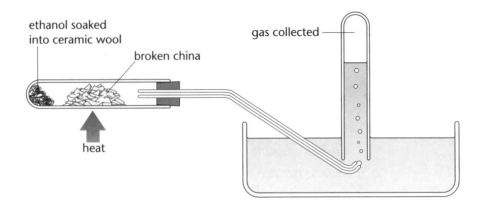

The gas collected was tested. The results are shown in the table.

Test	Result
with lighted splint	burns
with solution of bromine	turns from orange to colourless

a) Write down the name of the gas collected and draw its structure. ②

$$H-C=C-H$$ (structure drawn: $\underset{H}{\overset{H}{\diagup}}C=C\underset{H}{\overset{H}{\diagdown}}$)

b) Write a symbol equation for the reaction taking place at the broken china. ②

$$C_2H_5OH \rightarrow H_2O + C_2H_4$$

c) The alcohol propanol has a formula C_3H_7OH.

   i) The alcohol can exist in two isomers.
   Draw the structure of the two isomers. ②

   (structures drawn: $H-C-C-C-O-H$ and $H-C-C-C-H$ with $O-H$ group)

   ii) Dehydration of either isomer produces the same product.
   Draw the structure of the product. ②

   (structure drawn: $H-C-C=C$)

TOTAL 8

2 Ethanol, C_2H_5OH, can be produced from ethene, C_2H_4.

Alternatively, it can be made from glucose solution.

a) The reaction of ethene to produce ethanol is an addition reaction.

  i) What is an addition reaction?

      reaction when 2 ~~products~~ reactants are reacted to form 1 product ①

  ii) Write a balanced equation for the reaction.

      $C_2H_4 + H_2O \rightarrow C_2H_5OH$ ①

b) i) Write a balanced symbol equation for the formation of ethanol from glucose solution.

      $C_6H_{12}O_6 \rightarrow 2C_2H_5OH + 2CO_2$ ③

  ii) Describe how aqueous ethanol can be prepared from glucose solution.

      ⑤+①

  iii) How can a solution of ethanol be concentrated?
      Explain what this method of separation depends upon.

      Fractional distillation.

      melting point difference between water & ethanol ②

c) Explain why in some parts of the world ethanol is produced from ethene but in other parts it is made from sugar.

      ②

TOTAL 15

3 Ethyl methanoate is an ester, it has the structure:

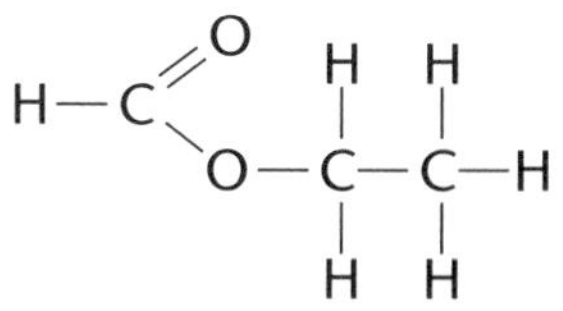

a) Write down the names of two substances that react together to form ethyl methanoate.

      methanoic acid and *ethanol* ②

b) Write down the other substance produced.

      ①

c) What conditions are needed to produce an ester?

      ②

d)   Many natural fats and oils are esters.

  i)   How are these oils converted into margarine?

....

.... ②

  ii) What are the products formed when a natural fat or oil is split up by hydrolysis?

....and.. ②

  iii) What conditions are needed for the hydrolysis of esters?

....

.... ②

TOTAL 11

④   Ethyne is a gas which belongs to the homologous series called alkynes.

a)   What is the meaning of the term homologous series?

....

.... ②

The structure of ethyne is:

$$H-C\equiv C-H$$

b)   Ethyne was once called acetylene. It is used for welding and cutting metals.
     Write a balanced symbol equation for the complete combustion of ethyne.

.... ③

c)   In 2000 three chemists received a Nobel prize for producing a conducting polymer called poly(ethyne). It still contains a double bond between carbon atoms.

  i)   Suggest a use for this polymer and a reason why it is better than metals for this use.

....

.... ②

  ii) Suggest a structure of this polymer.   ①

TOTAL 8

⑤   A hydrocarbon X contains 82.7% carbon.

a)   Calculate the empirical (simplest) formula of X.
     (Relative atomic masses: C = 12, H = 1)

....C..........H..

....82.7.......17.3..

....6.89.......17.3..

....I...........2.5..

....$= C_2H_5$..

.... ④

b) The relative formula mass of X is 58.

What is the molecular formula of X?

...........*C₄H₁₀₄₄*.. ②

c) What homologous series does X belong to?

........*et alkane*... ①

d) X can exist in two isomers.

i) What is meant by the term isomer?

......*same ~~state~~ formula different structure.*.. ②

ii) Draw the structures of the two isomers. ②

TOTAL 11

6 Polymers are widely used today.

a) What problems are caused in the disposal of polymers?

....

....

....

....

.... ⑤

b) Despite these problems industry continues to make new polymers rather than try to recycle existing ones. Explain why.

....

.... ②

TOTAL 7

7 Ethanoic acid is a weak acid with a molecular formula $C_2H_4O_2$.

a) Draw the structural formula of ethanoic acid. Put a ring around the replaceable hydrogen in the structure.

③

b) Ethanoic acid is produced when ethanol vapour and air are passed over a heated catalyst. Write a balanced symbol equation for the reaction.

.... ③

TOTAL 6

Further carbon chemistry

1 a) Ethene ❶

$$H_2C = CH_2$$

(structure showing C=C double bond with two H on each carbon)

b) $C_2H_5OH \rightarrow C_2H_4 + H_2O$ ❷
(One mark for reactants and one for products.)

c) i)

(structure: H-C-C-C-OH chain with H atoms)

(structure: H-C-C-C-H chain with OH on middle carbon)
❷

ii)

(structure: H-C-C=C with H atoms — propene)

(Deduct one mark for each error.) ❷

2 a) i) A reaction between two reactants to produce a single product. ❶

ii) $C_2H_4 + H_2O \rightarrow C_2H_5OH$ ❶

b) i) $C_6H_{12}O_6 \rightarrow 2C_2H_5OH + 2CO_2$ ❸
(One mark for reactant, one for products and one for balancing.)

ii) Any five of the following points:
Add yeast
Correctly named enzyme – zymase
In absence of air or oxygen
Temperature 35–40ºC
Process called fermentation
Stops when about 13% ethanol
Reaction slightly exothermic
(Plus one mark for Quality of Written Communication. This is for using correct scientific terminology.) ❺+❶

iii) (Fractional) distillation ❶
Ethanol and water have different boiling points ❶

c) In some areas (e.g. western Europe) where there are large amounts of ethene from cracking hydrocarbons, ethanol is made from ethene. ❶
In other areas where there are not supplies of ethene but sugar or other carbohydrates and sunlight, fermentation is used. ❶

3 a) Methanoic acid and ethanol. ❷

b) Water ❶

c) Heat, mixture of acid and alcohol ❶
with concentrated sulphuric (or phosphoric) acid ❶

d) i) Mixture of oil and hydrogen passed over heated nickel catalyst ❷

ii) Glycerol ❶
fatty acid ❶

iii) Heat with acid or alkali ❷

4 a) A family of compounds with similar properties and the same general formula. ❷

b) $2C_2H_2 + 5O_2 \rightarrow 4CO_2 + 2H_2O$ ❸

c) i) Replacing metals in electrical wiring ❶
Shortage of metals makes them expensive/polymers have lower density ❶

ii)

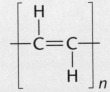

❶

EXAMINER'S TIP

You are not expected to have studied poly(ethyne) but to apply your knowledge to a new situation.

5   A hydrocarbon X contains 82.7% carbon.

a)   X contains (100 – 82.7) = 17.3% hydrogen ❶

              C   H

   Divide by RAM $\underline{\dfrac{82.7}{12}}$ $\underline{\dfrac{17.3}{1}}$

                6.9   17.3   ❶

   Ratio   1   2.5   ❶

   Empirical formula = C_2H_5 ❶

b)   Mass of 1 mole of C_2H_5 = 29   ❶

     Molecular formula = C_4H_{10} ❶

c)   Alkane   ❶

d) i)   Compounds with the same molecular formula   ❶

      but different structures.   ❶

EXAMINER'S TIP

Do not confuse the term isotope with the terms isomer and allotrope.

ii)

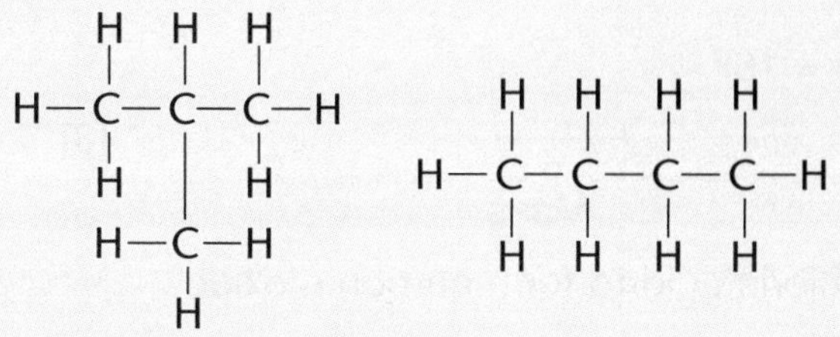

❷

6 a)   Dumping polymers in a landfill site   ❶

     Polymers are not biodegradable/do not rot away   ❶

     Burning polymers   ❶

     Can produce carbon monoxide   ❶

     Can produce acidic gases, e.g. HCl   ❶

b)   Difficult to sort into different types of polymer   ❶

     Cheaper to produce new than recycle   ❶

7 a)

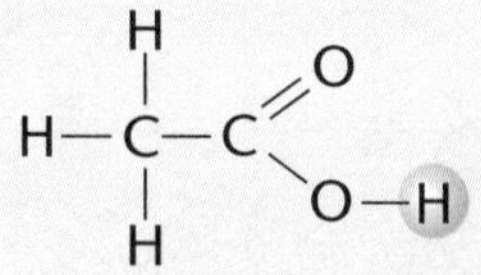

❸

EXAMINER'S TIP

There are 2 marks for the structure and 1 for circling the correct hydrogen.

b) $C_2H_5OH + O_2 \rightarrow CH_3COOH + H_2O$ ❸

CHAPTER 8

Quantitative chemistry

To revise this topic more thoroughly, see Chapter 8 in *Letts Revise GCSE Chemistry Study Guide*.

 Try this sample GCSE question and then compare your answers with the Grade C and Grade A model answers on the next page.

This question is about two reactions of calcium carbonate.

a   Calcium carbonate decomposes on heating to produce calcium oxide and carbon dioxide.

  (i)   Write a balanced symbol equation for this reaction.

  .... **[2]**

  (ii)   Calculate the mass of carbon dioxide formed when 0.1 moles of calcium carbonate is completely decomposed.

  (Relative atomic masses: Ca = 40, C = 12, O = 16)

  [3]

  Mass = ...g

b   Calcium carbonate reacts with dilute hydrochloric acid to form carbon dioxide.

  The equation for this reaction is shown below.

  $$CaCO_3(s) + 2HCl(aq) \rightarrow CaCl_2(aq) + CO_2(g) + H_2O(l)$$

  (i)   How many moles of hydrochloric acid would react with 1 mole of calcium carbonate?

  .... **[1]**

  (ii)   How many moles of hydrochloric acid would react with 0.1 moles of calcium carbonate?

  .... **[1]**

  (iii)   What volume of hydrochloric acid (1 mol/dm³) would contain 2 moles of hydrochloric acid?

  .... **[1]**

  (iv)   What volume of hydrochloric acid (1 mol/dm³) would react exactly with 0.1 moles of calcium carbonate?

  .... **[1]**

  (v)   How many moles of carbon dioxide would be produced when 0.1 moles of calcium carbonate completely react?

  .... **[1]**

  (vi)   What volume of carbon dioxide would be produced when 0.1 moles of calcium carbonate completely react?

  (1 mole of any gas at room temperature and pressure occupies 24 dm³)

  .... **[1]**

(Total 11 marks)

These two answers are at grades C and A. Compare which one your answer is closest to and think how you could have improved it.

GRADE C ANSWER

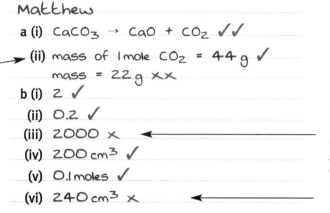

In this answer the examiner can award the first mark but there is no more working. The answer is wrong and the examiner cannot trace back and find the error. No further marks can be given.

Matthew

a (i) $CaCO_3 \rightarrow CaO + CO_2$ ✓✓

(ii) mass of 1 mole CO_2 = 44 g ✓
mass = 22 g ✗✗

b (i) 2 ✓

(ii) 0.2 ✓

(iii) 2000 ✗

(iv) 200 cm³ ✓

(v) 0.1 moles ✓

(vi) 240 cm³ ✗

Here the answer is correct but there are no units. The mark is given for the volume and the units.

This answer is incorrect.

7 marks = Grade C Answer

Grade booster ····⟩ move a C to a B
Always show your working so the examiner can go back and award method marks.
Always give the correct units.

GRADE A ANSWER

An excellent attempt at the question. Clear working out in here makes it easy for the examiner to follow.

Sanjay

a (i) $CaCO_3 \rightarrow CaO + CO_2$ ✓✓

(ii) 1 mole of $CaCO_3$ produces 1 mole of CO_2 ✓
44 g of CO_2 ✓
0.1 moles produces   4.4 g of CO_2 ✓

b (i) 2 ✓

(ii) 0.2 ✓

(iii) 2000 cm³ ✓

(iv) 200 cm³ ✓

(v) 0.1 moles ✓

(vi) 2.4 dm³ ✓

11 marks = Grade A* Answer

Grade booster ····⟩ move A to A*
This is an excellent answer that could not be improved.

Additional questions on quantitative electrolysis will be found in Chapter 9.

For Double Award Science papers calculations often involve relative formula masses. On Chemistry papers questions may be asked in terms of moles.

Students frequently do not understand the concept of moles as an amount of substance.

For further information about the mole refer to pages 124–125 of *Revise GCSE Chemistry Study Guide*.

Quantitative chemistry

1 Washing soda is hydrated sodium carbonate, $Na_2CO_3.xH_2O$.

Tricia carried out an experiment to find a value for x. She knows that when hydrated sodium carbonate is heated it loses all of its water and anhydrous sodium carbonate remains.

a) Describe, in full, the experimental method she should use.

....

....

....

....

....

....

....

....

.... ④+①

b) How can she work out the value of x from her results?

....

....

....

.... ④

TOTAL 9

2 Washing soda is hydrated sodium carbonate, $Na_2CO_3.xH_2O$.

Sam carried out an experiment to find the value of x.

28.6 g of fresh hydrated sodium carbonate crystals were dissolved in distilled water and the solution made up to a volume of 1000 cm^3.

25 cm^3 of this solution was transferred to a conical flask and a few drops of methyl orange indicator added.

The solution in the conical flask was titrated with hydrochloric acid (0.25 mol/dm^3) and 20.0 cm^3 of acid were required to reach the end point.

a) Write down the name of the most suitable piece of apparatus for measuring out exactly 25 cm^3 of sodium carbonate solution and transferring it to the conical flask.

.... ①

b) How would you know that the end point of the titration had been reached?

.... ①

c) Write a balanced symbol equation for the reaction of sodium carbonate with dilute hydrochloric acid.

.... ③

d) How many moles of hydrochloric acid react with $25\ cm^3$ of sodium carbonate solution?

Number of moles = ...

e) How many moles of hydrochloric acid would react with $100\ cm^3$ of sodium carbonate solution?

....

....

.... ②

Number of moles = ...

f) Using the equation in c), how many moles of anhydrous sodium carbonate are present in $1000\ cm^3$ of sodium carbonate solution?

②

Number of moles = ...

g) What mass of anhydrous sodium carbonate is present in $1000\ cm^3$ of sodium carbonate solution? (Relative atomic masses: Na = 23, C = 12, O = 16)

②

Mass = ... g

h) Complete the following:

.... g of anhydrous sodium carbonate combines with

.... g of water to produce 28.6 g of sodium carbonate crystals. ①

i) How many grams of water would combine with 1 mole of sodium carbonate?

.... ①

j) What is the value of x? (Relative atomic masses: H = 1, O = 16)

.... ①

TOTAL 16

3 Urea, $CO(NH_2)_2$, and ammonium nitrate, NH_4NO_3, are two nitrogen fertilisers.
Ammonium nitrate is very soluble in water but urea is only slightly soluble.

a) Calculate the percentage of nitrogen in:

i) urea

③

Percentage ....%

ii) ammonium nitrate.

③

Percentage .... %

b)   Suggest **two** advantages of urea over ammonium nitrate.

   i)   ....

   ii) .... ②

c)   What happens when excess nitrogen fertilisers are washed into rivers?

   ....

   ....

   ....

   .... ③+①

TOTAL 12

---

4   There are two chlorides of mercury. One of these mercury chlorides is reduced to mercury in an experiment.

   Here are the results:

   Mass of beaker = 45.23 g

   Mass of beaker + mercury chloride = 47.94 g

   Mass of beaker + mercury produced = 47.23 g

a)   From these results calculate the mass of mercury chloride used and mercury produced. ②

   Mass of mercury chloride   = ... g

   Mass of mercury produced = ... g

b)   Complete the following:

   .... g of mercury combines with .... g of chlorine to produce

   .... g of mercury chloride. ①

c)   How many moles of mercury atoms were produced in this experiment?

   (Relative atomic mass; Hg = 200)

                                          Number of moles = ... ②

d)   How many moles of chlorine atoms were combined with mercury in the sample of mercury chloride? (Relative atomic mass: Cl = 35.5)

                                          Number of moles = ... ②

e)   How many moles of chlorine atoms would combine with 1 mole of mercury atoms?

   .... ①

f)   What is the formula of the mercury chloride used?

   .... ①

TOTAL 9

5 Ali carried out an experiment to find the formula of copper bromide.

21.6 g of copper bromide was heated with an excess of iron powder.

After the reaction, 9.6 g of copper were formed.

a) What mass of bromine was in the sample of copper bromide?

Mass = ...g ②

b) How many moles of bromine atoms were in the sample of copper bromide?

(Relative atomic masses: Br = 80)

Number of moles = ...g ②

c) How many moles of copper atoms were formed?

(Relative atomic masses: Cu = 64)

Number of moles = ...g ②

d) What is the simplest formula of the copper bromide?

.... ①

e) Write a balanced symbol equation for the reaction between iron and copper bromide.

.... ③

TOTAL 10

---

6 Ammonia burns in excess oxygen to produce nitrogen and water.

The equation for the reaction is shown below.

$$4NH_3(g) + 3O_2(g) \rightarrow 2N_2(g) + 6H_2O(l)$$

40 cm^3 of ammonia was burned using 50 cm^3 of oxygen.

Assume all volumes are measured at room temperature and pressure.

a) What volume of oxygen reacts with 40 cm^3 of ammonia?

Volume = ... cm^3 ①

b) What volume of oxygen was unreacted?

Volume = ... cm^3 ①

c) What is the total volume of gas remaining?

Volume = ... cm^3 ①

TOTAL 3

QUESTION BANK ANSWERS

❶ a) Any four points from:
weigh evaporating basin or suitable container
weigh container and sample of washing soda crystals
heat the container and crystals
cool and re-weigh
reheat the container
re-weigh and repeat until constant mass
(In addition one mark is available for Quality of Written Communication. For this your answer has to be written so the events are in a correct and logical order.) **❹+❶**

b) Find the mass of water lost **❶**
Find the mass of anhydrous salt **❶**
Find the mass of water combined with 1 mole of Na_2CO_3 (106 g) **❶**
Divide this mass by the mass of 1 mole of water (18 g) to find the value of x. **❶**

EXAMINER'S TIP

Although this question does not involve numbers it is about the quantitative method. Candidates often find it more difficult to explain what to do than just handle numbers.

❷ a) Pipette (measuring cylinder is not accurate enough) **❶**
b) Indicator (methyl orange) changes colour **❶**
c)
$Na_2CO_3(aq) + 2HCl(aq) \rightarrow 2NaCl(aq) + CO_2(g) + H_2O(l)$ **❸**
(One mark for correct formulae of reactants, one for correct formulae of products and one mark for balancing. State symbols are optional as they are not expressly asked for.)

EXAMINER'S TIP

Phenolphthalein is another indicator but it changes colour at a higher pH when the following reaction takes place.
$Na_2CO_3(aq) + HCl(aq) \rightarrow NaHCO_3(aq) + NaCl(aq)$
The volume of acid required for this process is half the volume required when using methyl orange as indicator.

d) 20.0 cm³ of hydrochloric acid (0.25 mol/dm³)
Number of moles in 1000 cm³ = 0.25
Number of moles in 20 cm³ of acid = 0.25 × 20/1000 **❶**
= 0.005 moles **❶**
e) 0.005 moles of acid reacts with 25 cm³ of sodium carbonate
0.005 × 1000/25 moles of acid react with 1000 cm³ sodium carbonate solution **❶**
= 0.2 moles **❶**
f) From the equation, 2 moles of acid react with 1 mole of sodium carbonate. **❶**
Number of moles of sodium carbonate = 0.1 moles **❶**
g) Mass of sodium carbonate = 0.1 × $M_r(Na_2CO_3)$ **❶**
= 10.6 g **❶**
h) 10.6 and 18.0 g **❶**
i) 180 g **❶**
j) 10 **❶**

EXAMINER'S TIP

This is a difficult question but splitting it up into small steps makes it easier.

❸ a) i) Urea
Mass of 1 mole of urea = 60 g **❶**
Mass of nitrogen in each mole = 28 g **❶**
Percentage = $\frac{28}{60}$ × 100 = 47% **❶**
ii) Ammonium nitrate
Mass of 1 mole of ammonium nitrate = 80 g **❶**
Mass of nitrogen in each mole = 28 g **❶**
Percentage = $\frac{28}{80}$ × 100 = 35% **❶**
b) Lasts longer in the soil. **❶**
Higher percentage of nitrogen. **❶**
c) Any three of the following:
bacteria in river oxidise ammonia into nitrates
reduces dissolved oxygen concentration in water
nitrogen encourages plant growth
this prevents light entering/reduces photosynthesis
decaying plants use up oxygen
fish die **❸+❶**
(Plus one mark for Quality of Written Communication. This is awarded if the steps are in a logical order and communicate the Science well.)

4 a) 2.71 g   **❶**
  2.00 g   **❶**

b) 2.00 g of mercury combines with 0.71 g of chlorine to produce 2.71 g of mercury chloride.   **❶**

c) 0.01   **❷**
d) 0.02   **❷**
e) 2   **❶**
f) $HgCl_2$ **❶**

5 a) 21.6 – 9.6   **❶**
  12.0 g   **❶**

b) $\dfrac{12.0}{80}$ **❶**
  0.15   **❶**

c) $\dfrac{9.6}{64}$ **❶**
  0.15   **❶**

d) CuBr   **❶**

e) Two possible equations:
  $2CuBr + Fe \rightarrow 2Cu + FeBr_2$ or
  $3CuBr + Fe \rightarrow 3Cu + FeBr_3$ **❸**

---

6 a) 30 cm³   **❶**

EXAMINER'S TIP

From the equation, 4 volumes of ammonia react with 3 volumes of oxygen.

b) 20 cm³   **❶**

EXAMINER'S TIP

30 cm³ reacted so 20 cm³ remained.

c) 40 cm³   **❶**

EXAMINER'S TIP

20 cm³ nitrogen produced plus 20 cm³ oxygen. Total volume = 40 cm³
Don't forget the water has condensed so the volume is negligible.

Quantitative chemistry

CHAPTER 9

Electrolysis

To revise this topic more thoroughly, see Chapter 9 in *Letts Revise GCSE Chemistry Study Guide.*

Try this sample GCSE question and then compare your answers with the Grade C and Grade A model answers on the next page.

The diagram shows an apparatus that can be used to investigate the effect of electricity on various substances.

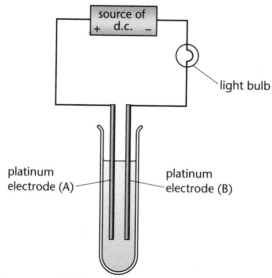

a What will be seen if the boiling tube contains distilled water and some sodium chloride is added?

....

.... **[2]**

b Write down two ways of speeding up the rate of electrolysis of a solution using this apparatus. You should not change the concentration of the solution.

(i) ....

(ii) .... **[2]**

c In industry, during the electrolysis of concentrated sodium chloride, hydrogen is made at the steel cathode and chlorine is produced at the titanium anode.

(i) What is the other product of this process?

.... **[1]**

(ii) Write an ionic equation for the reaction at the cathode.

.... **[3]**

(iii) Why should the anode not be made of iron?

.... **[2]**

(iv) How would the products be different if a dilute solution of sodium chloride was used?

.... **[1]**

(Total 11 marks)

These two answers are at grades C and A. Compare which one your answer is closest to and think how you could have improved it.

GRADE C ANSWER

Gita only gives one answer. As there are two marks available she should be looking for a second answer.

H₂ instead of 2H would have given a completely correct ionic equation.

Gita

a   Bulb lights ✓

b   Raise the temperature ✗ Add a catalyst ✗

c (i)   Sodium hydroxide ✓

   (ii) $2H^+ + 2e^- \rightarrow 2H$ ✓ ✗ ✓

   (iii)   Iron cathode might react with chlorine ✓ ✗

   (iv)   Oxygen produced ✓

Gita is confused with rate of reaction. She gives two things that change rate of reaction.

The name of the product iron(III) chloride should have been given.

6 marks = Grade C Answer

Grade booster ····❯ move a C to a B
Gita's answer lacks detail in places. Where two marks are awarded two points must be made.

GRADE A ANSWER

The answer here is good, the examiner would have awarded the mark just for mentioning bubbles.

Another possible answer here would be to push the electrodes deeper into the boiling tube. This effectively increases the surface area of the electrodes.

Kim

a   Bulb lights up ✓ Bubbles of gas formed at electrodes ✓

b   Increase voltage of power source ✓ Move electrodes closer together ✓

c (i)   Sodium hydroxide ✓

   (ii) $2H^+ + 2e^- \rightarrow H_2$ ✓✓✓

   (iii)   Iron may corrode ✗ ✗

   (iv)   Oxygen as low concentration of chloride ions causes hydroxide ions to be discharged. ✓

This answer is disappointing. Kim just links iron with rusting and does not consider the situation in the question.

The answer is more than expected.

9 marks = Grade A Answer

Grade booster ····❯ move A to A*
This is a very good answer. Only in one part, c (iii), is the answer wrong. Kim has not looked at the situation in the question – iron in contact with chlorine.

Electrolysis

85

Electrolysis

1 This question is about electrolysis of water.

Some sulphuric acid is added to the water to make it conduct electricity.

The solution contains H^+, OH^- and SO_4^{2-} ions.

a) Which ion will be discharged at the negative electrode (cathode)?

.... ①

b) Which gas is produced at the negative electrode?

.... ①

c) Oxygen gas is produced at the positive electrode. Finish the ionic equation:

.... $\rightarrow O_2 + H_2O + e^-$ ②

d) The overall equation for the electrolysis of water is

$$2H_2O \rightarrow 2H_2 + O_2$$

During the electrolysis, $50\,cm^3$ of hydrogen is collected.

What volume of oxygen is collected in the same period?

Explain your answer.

....

.... ②

e) With copper electrodes, a blue colour is seen around the positive electrode.

i) Which ion present in the solution causes this blue colour?

.... ①

ii) Finish the ionic equation

$Cu \rightarrow$

②

TOTAL 9

2 The table gives the products of electrolysis of some solutions.

Solution	Product at negative electrode	Product at positive electrode
dilute sulphuric acid	hydrogen	oxygen
copper(II) sulphate	copper	oxygen
silver nitrate	silver	oxygen
sodium nitrate	hydrogen	oxygen
copper(II) nitrate		
dilute nitric acid		

a) Complete the table by filling in the four boxes. ④

b)   At which electrode is a metal or hydrogen produced?

.... ①

c)   The electrolysis of zinc bromide forms zinc and bromine.

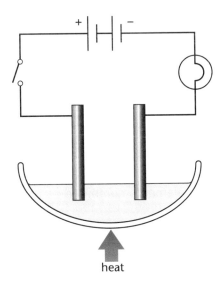

heat

i)   On the diagram, label the anode (positive electrode) and the cathode
     (negative electrode).   ①

ii) Write down the name of the electrolyte.. ①

iii) What must happen before the electrolysis begins?

     .... ①

iv) Write down two things you would see that show electrolysis is taking place.

     ....

     .... ②

v)   During the experiment the bulb suddenly shines brighter and continues

     to shine even when the heat is removed and the zinc bromide solidifies.

     Suggest a reason for this.

     ....

     ....

     .... ③

     TOTAL 13

3 The table gives the colours of six salts.

Salt	Colour
copper(II) sulphate	blue
copper(II) nitrate	blue
potassium chromate	yellow
sodium chromate	yellow
sodium nitrate	colourless
potassium nitrate	colourless

a) Write down the names of the three elements combined in potassium chromate.

.... ②

b) Suggest the colour of:

i) sodium ions ....

ii) copper(II) ions ....

iii) chromate ions..

iv) a solution of copper(II) chromate .... ④

c) The apparatus shown in the diagram was set up.

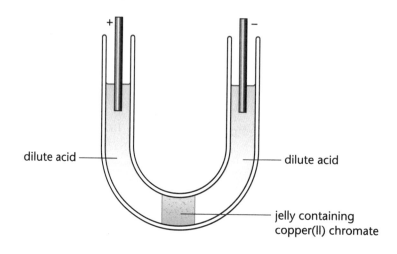

dilute acid

dilute acid

jelly containing
copper(II) chromate

After some time a yellow band moves through the gel towards the anode and a blue band towards the cathode. Explain these observations.

....

....

....

.... ④

TOTAL 10

4 In the nickel plating of a metal object, the object is made the negative electrode in the cell containing nickel(II) sulphate.

A current of 19.2 A is passed through the cell for 100 seconds.

a) How many coulombs were passed? ①

Answer = ... C

b) How many faradays of electricity were passed? ①

(1 faraday = 96 000 C)

Answer = ... F

c) How many moles of nickel atoms should have been deposited on the electrode in this time? (Charge on nickel ion is +2)

.... ①

d) What should be the increase in mass of the object?

(Relative atomic mass: Ni = 59)

.... ②

e) Suggest why the change in mass is less than expected.

.... ①

TOTAL 6

Electrolysis

1 a) H^+ ❶
 b) H_2 ❶
 c) $4OH^- \rightarrow O_2 + 2H_2O + 4e^-$ ❷
 (One mark for reactant and one mark for balancing.)
 d) $25\,cm^3$ ❶
 From equation, 2 volumes of hydrogen produced to 1 volume of oxygen ❶
 e) i) Cu^{2+} ❶
 ii) $Cu \rightarrow Cu^{2+} + 2e^-$ ❷
 (One mark for products and one mark for balancing.)

2 a) copper   oxygen
 hydrogen   oxygen ❹
 b) Negative electrode ❶

EXAMINER'S TIP

Part b) is answered from the table.

(c)i)

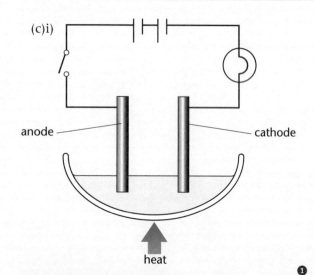

anode   cathode

heat ❶

EXAMINER'S TIP

One mark is for labelling the electrodes and one mark for getting them the right way round. The anode (positive electrode) is connected to the positive terminal of the power source.

ii) Zinc bromide ❶
iii) Zinc bromide must be melted ❶
iv) Bulb lights ❶
 Brown gas (bromine) escaping ❶

v) Zinc produced connects anode and cathode ❶
 Electricity conducts through metal ❶
 Better than through the melt ❶

EXAMINER'S TIP

The third point in c) v) explains why the bulb glows brighter.

3 a) Potassium, chromium, oxygen ❷
 (Two marks for three correct names – one mark for two correct.)

EXAMINER'S TIP

A common mistake here is to give the name chrome instead of chromium. If you are in doubt, look at the Periodic Table.

b) i) colourless ❶
 ii) blue ❶
 iii) yellow ❶
 iv) green ❶
 c) Blue copper(II) ions move ❶
 towards negative rod ❶
 Yellow chromate ions move ❶
 towards positive rod ❶

EXAMINER'S TIP

Ions migrate towards the electrode of opposite charge.

4 a) 1920 ❶
 b) $\frac{1920}{96\,000} = 0.02$ ❶
 c) 0.01 ❶
 d) 0.01 × 59 ❶
 = 0.59 g (units essential) ❶
 e) Some nickel flakes off ❶

Centre number
Candidate number
Surname and initials

 **Examining Group**

General Certificate of Secondary Education

Chemistry
Paper 1
Higher tier

Time: one and a half hours

Instructions to candidates

Write your name, centre number and candidate number in the boxes at the top of this page.

Answer ALL questions in the spaces provided on the question paper.

Show all stages in any calculations and state the units. You may use a calculator.

Include diagrams in your answers where this may be helpful.

Information for candidates

The number of marks available is given in brackets **[2]** at the end of each question or part question.

The marks allocated and the spaces provided for your answers are a good indication of the length of answer required.

For Examiner's use only	
1	
2	
3	
4	
5	
6	
7	
Total	

EDUCATIONAL

1 Magnesium reacts with oxygen to form solid magnesium oxide.

(a) Describe the changes in the electron arrangement of the atoms that occur when magnesium burns in oxygen.

The 2 outer electron in Mg will be transferred to the outer shell of the oxygen. Mg lose 2e⁻ (oxidation) O₂ gain e 2e⁻ (Reduction) **[3]**

(b) Five groups of students burn different masses of magnesium in air using the apparatus in the diagram.

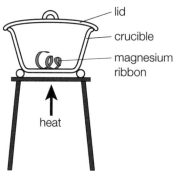

lid
crucible
magnesium
ribbon

heat

Here are their results.

Group	A	B	C	D	E
mass of magnesium in g	0.20	0.30	0.40	0.50	0.60
mass of magnesium oxide in g	0.33	0.50	0.63	0.88	1.00
mass of oxygen combined with magnesium in g	0.13	0.20	0.23	0.38	0.40

(i) Draw a graph of the mass of magnesium (on the x-axis) against the mass of oxygen.

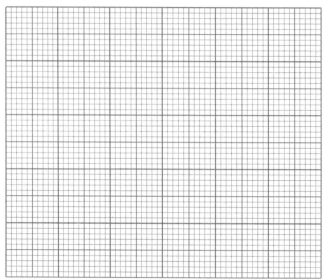

[3]

Letts

(ii) Which group is most likely to have unburned magnesium in the crucible at the end of the experiment? Explain your choice.

Group ....

Explanation .... **[2]**

(iii) Use the information in the graph to show that the formula of magnesium oxide is MgO. (Relative atomic masses: O = 16, Mg = 24)
You must show how you used the graph and all your working.

....

....

.... **[3]**

(c) **(i)** One of the students finds out that magnesium also reacts with nitrogen. Write down the name of the compound formed.

.... **[1]**

(ii) Air contains four times as much nitrogen as oxygen by volume.
Suggest why, despite this, you would not expect much of the compound of magnesium and nitrogen to be formed in this experiment.

.... **[1]**

(d) Two hundred and fifty years ago scientists had a different way of explaining combustion.

They believed that:

• when something burned it lost a substance called phlogiston.

• a substance that did not burn contained no phlogiston

Explain how this theory lost support quickly when accurate balances were made.

....

.... **[2]**

(Total 15 marks)

[turn over

Letts

2 **(a)** The diagram shows a 'dot and cross' diagram for a hydrogen chloride molecule.

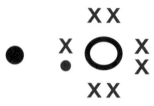

  (i) What do each of the following represent?

  ● ....Hydrogen....ion........... ○ ....Chloride ion...........

  X ....e⁻ from Chloride........... ● ....e⁻ from Hydrogen........... **[4]**

  (ii) What type of bonding is shown in a molecule of hydrogen chloride?

  ....covalent........... **[1]**

(b) When hydrogen chloride is dissolved in water, each molecule is split into two ions.

  (i) Draw 'dot and cross' diagrams to show the two ions formed.

  [2]

  (ii) Describe a test to show that a solution of hydrogen chloride in water contains ions.

  ....

  .... **[2]**

  (iii) Describe a test using an indicator to show that a solution of hydrogen chloride in water contains H^+ ions.

  ....

  .... **[2]**

Letts

Leave blank

(c) When a solution of hydrogen chloride is added to a carbonate, bubbles of carbon dioxide gas are produced.

(i) Describe a test for carbon dioxide.

....

.... **[2]**

(ii) Complete an ionic equation for this reaction.

CO_3^{2-} + ... → CO_2 + ... **[2]**

(Total 15 marks)

[turn over

3   (a) Magnesium reacts with dilute hydrochloric acid to produce hydrogen gas and magnesium chloride solution.
Write a symbol equation for the reaction.

.... [3]

   (b) How would you show that the gas is hydrogen?

....

.... [2]

   (c) The apparatus in the diagram is used to study the rate of reaction between magnesium and dilute hydrochloric acid.
The reaction is carried out at 15°C.

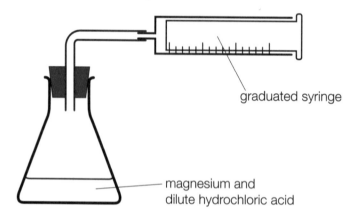

graduated syringe

magnesium and
dilute hydrochloric acid

The results are shown in the table.

Time in seconds	Volume of gas in cm^3
0	0
10	20
20	30
30	39
40	40
60	44
80	45
100	45

Plot the results on the grid. Draw the best line through the points.

[3]

(d) Sally predicts that if the reaction was repeated at 30°C the reaction would be faster.

(i) How would the graph be different from the graph in **(c)**?

.... [1]

(ii) Explain why the reaction is faster. Use ideas of particles in your answer.

....

....

....

.... [3]

(e) If $5\,cm^3$ of copper(II) sulphate is added to the hydrochloric acid before adding the magnesium, the reaction is faster.
Suggest an explanation for this observation. Include an equation in your answer.

....

....

....

....

.... [5]

(Total 17 marks)

[turn over

Letts

4   Titanium(IV) oxide, TiO_2, is found as a mineral in the ore rutile.
Rutile is heated with carbon and chlorine gas to produce titanium(IV) chloride and carbon monoxide.

(a) Write a balanced symbol equation, with state symbols, for the reaction producing titanium(IV) chloride.

.... **[4]**

(b) How can titanium(IV) chloride be separated from the remains of the rock?

....

.... **[2]**

Titanium is extracted by heating titanium(V) chloride with sodium.

(c) Write a balanced symbol equation for the reaction of titanium(IV) chloride and sodium.

.... **[3]**

(d) Why is titanium an expensive metal, despite the fact that it is the seventh most common metal in the Earth?

.... **[1]**

(Total 10 marks)

5 Diamond and graphite are two allotropes of carbon.

Leave blank

(a) Explain the meaning of the term allotrope.

....

.... **[1]**

(b) Describe the structure of diamond and graphite.

Diamond

....

....

Graphite

....

.... **[4]**

(c) Write down one property of each allotrope and explain how the structure of the allotrope is consistent with this property.

Diamond – property ....

Explanation ....

....

Graphite – property ....

Explanation ....

.... **[4]**

(d) CFCs are compounds of carbon, chlorine and fluorine. One such compound is dichlorodifluoromethane, CF_2Cl_2. This has a melting point of $-158°C$ and a boiling point of $-30°C$.

(i) Is dichlorodifluoromethane a solid, liquid or gas at room temperature and atmospheric pressure?

.... **[1]**

99

[turn over

Letts

(ii) What type of structure does dichlorodifluoromethane have?

.... **[1]**

(iii) What type of bonding is present in dichlorodifluoromethane?

.... **[1]**

(Total 12 marks)

6 The table gives the names and formulae of the first five alkanes.

Name	Formula
methane	CH_4
ethane	C_2H_6
propane	C_3H_8
butane	C_4H_{10}

(a) Alkanes are **saturated hydrocarbons**.
Explain the meaning of the words in bold type.

....

.... **[2]**

(b) What is the molecular formula of an alkane containing *n* carbon atoms?

.... **[1]**

(c) Draw the structure of propane.

[1]

(d) Propane is used as a fuel in camping gas.

    (i) Write a balanced equation for the burning of propane in a plentiful supply of air.

.... **[3]**

    (ii) Draw an energy level diagram for the reaction.

[3]

    (iii) Why is it dangerous to have a limited supply of air when burning propane?

.... **[2]**

(Total 12 marks)

[turn over

Letts

7   The diagram shows a section of the Earth's crust.

A

B

D

C

X

(a) Which type of rock is found in layer A?

.... [1]

(b) Which type of rock is likely to be found in D?

.... [1]

(c) Which type of rock is likely to be found at X?

.... [1]

(d) Where would fossils not be found?

.... [1]

(e) How do you know that the rocks in C are older than the rocks in A?

.... [1]

(f) How do you know that the rocks in D are older than B and C but younger
than A?

....

.... [2]

(g) Describe the crystals likely to be found in D. Explain your answer.

....

.... [2]

(Total 9 marks)

102

 **Examining Group**

General Certificate of Secondary Education

For Examiner's use only	
1	
2	
3	
4	
5	
6	
Total	

Chemistry
Paper 2
Higher tier

Time: one hour

Instructions to candidates

Write your name, centre number and candidate number in the boxes at the top of this page.

Answer ALL questions in the spaces provided on the question paper.

Show all stages in any calculations and state the units. You may use a calculator.

Include diagrams in your answers where this may be helpful.

Information for candidates

The number of marks available is given in brackets **[2]** at the end of each question or part question.

The marks allocated and the spaces provided for your answers are a good indication of the length of answer required.

EDUCATIONAL

1 The solubility curve for potassium dichromate is shown below.

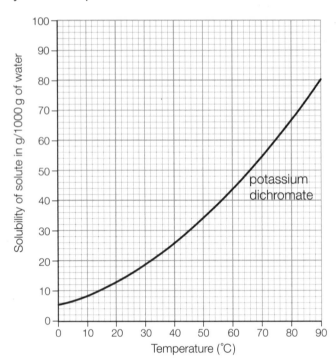

(a) How does the solubility of potassium dichromate change with increasing temperature?

.... **[1]**

Tim wanted to purify some potassium dichromate.
He added some potassium dichromate to 100 g of water and heated gently while stirring.
The compound just dissolved at 90°C.
The solution was then allowed to cool to 50°C.

(b) What mass of potassium dichromate was added (ignore impurities)?

.... **[1]**

(c) What mass of potassium dichromate would still be dissolved when the solution had cooled to 50°C?

.... **[1]**

(d) What mass of crystals would be produced after the solution had stopped crystallising at 50°C?

.... **[1]**

(e) How could Tim get a sample of dry crystals of potassium dichromate from this experiment?

.... **[2]**

(Total 6 marks)

2   Describe and explain tests that could be used to distinguish:

(a) solid sodium chloride and solid potassium chloride

....

.... [2]

(b) potassium chloride solution and potassium sulphate solution

....

....

.... [3]

(c) iron(II) sulphate solution and iron(III) sulphate solution

....

....

.... [3]

(d) ammonium chloride and magnesium chloride.

....

....

.... [3]

(Total 11 marks)

[turn over

3   The diagram shows the apparatus used by four groups of students in a
    quantitative experiment.

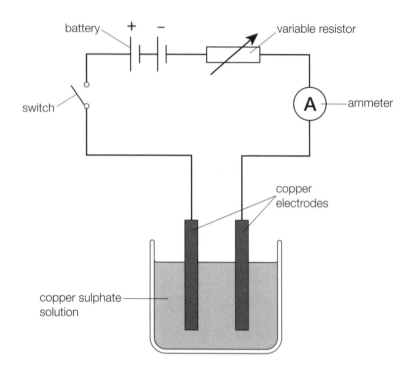

They found the loss of the mass of the anode in each experiment when a known
quantity of electricity was passed through the solution.

The table shows their results.

Group	Loss of mass of the anode/g	Quantity of electricity/ coulombs
1	0.015	45
2	0.040	120
3	0.076	230
4	0.096	290

(a) On the grid plot the mass of copper lost (on the y-axis) against the number of coulombs of electricity passed. Draw the best line.

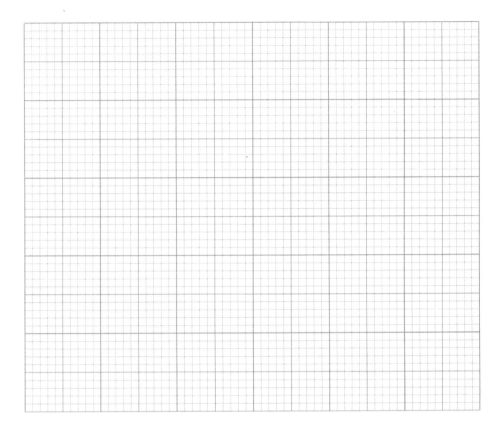

[3]

(b) (i) Use the graph to find how many coulombs would be required for 0.064 g of copper to be lost from the anode.

.... **[1]**

(ii) Calculate how many moles of electrons are needed for one mole of copper atoms to be lost. (1 faraday = 96 000C)
(Relative atomic mass: Cu = 64)

Number of moles = ... **[3]**

(c) Write an ionic equation, with state symbols, to show what happens to copper at the anode during the reaction.

.... **[4]**

(d) The students in group 4 used a current of 0.2 A. How long did the current pass? Give your answer to the nearest minute. **[3]**

Time = ... minutes

(Total 14 marks)

[turn over

4 The diagram shows the arrangement of atoms in pure iron.

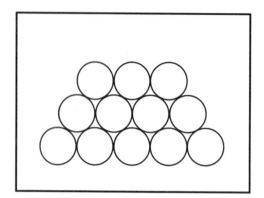

(a) Describe the arrangement of atoms in a layer of iron.

....

.... **[2]**

(b) Explain why iron is a good conductor of electricity.

....

.... **[2]**

(c) Add to the diagram to show the changes that take place when iron is turned into steel. **[1]**

(d) Use the diagram to explain why steel is harder than pure iron.

....

.... **[2]**

(Total 7 marks)

5   The table gives the names and formulae of some carbon compounds

Name	Formula
ethanol	C_2H_5OH
ethene	C_2H_4
butane	C_4H_{10}
butene	C_4H_8

(a) What type of reaction takes place when ethanol is converted into ethene?

.... **[1]**

(b) Ethanol is manufactured by two different methods – from ethene and from sugar.

(i)   Describe the method used to produce ethanol from ethene. Include an equation in your answer.

....

....

.... **[3]**

(ii) Describe the method used to produce ethanol from sugar. Include an equation in your answer.

....

....

....

....

.... **[5]**

(iii) Suggest why ethene is used in some parts of the world and sugar in others.

....

....

.... **[2]**

[turn over

Letts

(c) Butane can exist in two different structures.

    (i) What name is given to the existence of these two structures?

    .... **[1]**

    (ii) Draw the two structures.

    [2]

    (iii) Describe a chemical test to distinguish butane and butene.

    ....

    ....

    .... **[3]**

(Total 17 marks)

6   Chlorine is bubbled through a solution of potassium bromide.
    The equation for the reaction is shown below.

$$Cl_2 + 2KBr \rightarrow 2KCl + Br_2$$

(a) Describe the colour change you would see.

....

.... **[2]**

(b) Explain the terms oxidation and reduction, using this reaction.
     Your answer should involve electron transfer.

....

....

.... **[3]**

(Total 5 marks)

Letts

Answers to mock examination Paper 1

Question 1

(a) Magnesium 2,8,2   Oxygen 2,6
(Both required) **[1]**

Two electrons transferred from magnesium to oxygen. **[1]**

Both (ions) have an electron arrangement 2,8. **[1]**

(b) (i)

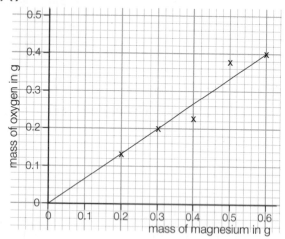

One mark for graph to fill over half grid with correct scales on axes, one mark for correct plotting, one mark for a straight line (with a ruler) through most of points. **[3]**

(ii) Group C **[1]**

Less oxygen combined than expected from graph. **[1]**

(iii) Values chosen from graph below 0.30 g **[1]**

Mass of magnesium divided by 24 and mass of oxygen divided 16. **[1]**

Comment that the two answers are the same or nearly the same. **[1]**

(c) (i) Magnesium nitride **[1]**

(ii) Nitrogen is the inactive gas in the air or oxygen is the active gas. **[1]**

(d) Any two points from:

combustion accompanied by an increase in mass;

providing all the products are collected;

these mass changes are very small and cannot be detected without an accurate balance;

an increase in mass is not consistent with the loss of a substance;

Unless the substance is given a negative mass. **[2]**

Total 15 marks

Question 2

(a) (i) Hydrogen nucleus, chlorine nucleus, electron from chlorine, electron from hydrogen. **[4]**

(ii) Covalent bonding **[1]**

(b) (i)

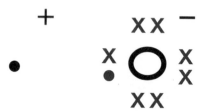

(One mark for each ion) **[2]**

(ii) Test electrical conductivity of solution **[1]**

Solution containing ions conducts electricity **[1]**

(iii) Named suitable indicator, e.g. litmus. **[1]**

Indicator turns to acid colour – red in the case of litmus. **[1]**

(c) (i) Test with limewater. **[1]**

Limewater turns milky **[1]**

(ii) $CO_3^{2-} + 2H^+ \rightarrow CO_2 + H_2O$ **[2]**

Total 15 marks

Question 3

(a) $Mg + 2HCl \rightarrow MgCl_2 + H_2$ **[3]**

(One mark for formulae of reactants, one mark for formulae of products and one mark for balancing.)

(b) Lighted splint **[1]**

Burns with a squeaky pop **[1]**

EXAMINER'S TIP

A frequent mistake here is to use a glowing splint. Obviously the candidate is confusing the test for hydrogen with the test for oxygen.

(c)

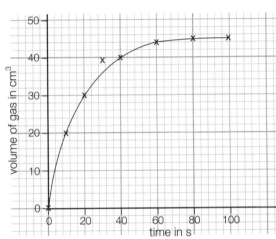

(One mark for correct scales and axes, one mark for correct plotting, one mark for best line.) **[3]**

EXAMINER'S TIP

There is an anomalous point at 30 seconds. Your best line should miss this point. Drawing graphs with anomalous points is intended for grades B and above.

(d) (i) The graph should be steeper than the one drawn in **(c)** **[1]**

EXAMINER'S TIP

The sketched graph would also finish at 45 cm³ at a time less than 80 s. The question does not ask about this.

(ii) Any three points from:

at a higher temperature acid particles move faster;

have greater average kinetic energies;

therefore more collisions between acid particles and metal;

greater number of collisions have sufficient energy for reaction. **[3]**

(e) Magnesium and copper(II) sulphate react **[1]**

Copper is produced **[1]**

Copper acts as a catalyst **[1]**

$Mg + CuSO_4 \rightarrow MgSO_4 + Cu$ **[2]**

EXAMINER'S TIP

Weaker candidates would write that copper(II) sulphate is a catalyst. This is worth 1 mark. Copper is an effective catalyst because it is deposited over the surface of the magnesium. It has a large surface area.

Total 17 marks

Question 4

(a) $TiO_2(s) + 2C(s) + 2Cl_2(g) \rightarrow TiCl_4(l) + 2CO(g)$ **[4]**

(One mark for formulae of reactants, one mark for formulae of products, one mark for balancing, one mark for correct state symbols throughout.)

(b) Titanium(IV) chloride escapes as vapour. **[1]**

Condenses to form liquid. **[1]**

(c) $TiCl_4 + 4Na \rightarrow Ti + 4NaCl$ **[3]**

(d) Sodium is an expensive metal/lot of energy required. **[1]**

Total 10 marks

Question 5

(a) Different forms of the same element in the same physical state. **[1]**

(b) Diamond – giant structure of carbon atoms; **[1]**

all strongly bonded together by covalent bonds. **[1]**

Graphite – carbon atoms are arranged in layers; **[1]**

only weak forces between the layers. **[1]**

EXAMINER'S TIP

Candidates are better at drawing the structures but sometimes will be asked to describe them. If you do draw them don't draw them too small.

(c) Diamond – very hard **[1]**

All covalent bonds are strong and must be broken **[1]**

Or

Poor conductor of electricity **[1]**

All electrons bound in covalent bonds. **[1]**

Graphite – slippery **[1]**

Layers slide over one another **[1]**

Or

Good conductor of electricity [1]

Free electrons throughout the structure can move and carry current. [1]

(d) (i) Gas [1]

  (ii) Molecular [1]

  (iii) Covalent [1]

Total 12 marks

Question 6

(a) Saturated. All the carbon–carbon bonds are single bonds. [1]

Hydrocarbon – compound of carbon and hydrogen only. [1]

(b) C_nH_{2n+2} [1]

(c)

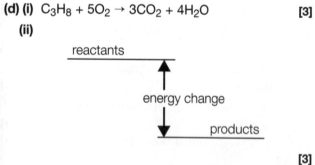

[1]

EXAMINER'S TIP

Don't miss off the hydrogen atoms and check you have a structure with three carbon atoms and eight hydrogen atoms.

(d) (i) $C_3H_8 + 5O_2 \rightarrow 3CO_2 + 4H_2O$ [3]

  (ii)

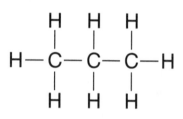

[3]

(One mark for reactants line above products line showing an exothermic reaction.

One mark for reactants and products labelled correctly.

One mark for energy change shown.)

  (iii) Carbon monoxide produced. [1]

Carbon monoxide is poisonous. [1]

Total 12 marks

Question 7

(a) Sedimentary [1]

(b) Igneous [1]

(c) Metamorphic [1]

(d) In igneous rock [1]

(e) The lower a layer of rock is in a sequence the older the rock. [1]

(f) D cuts into B and C but does not cut into A [1]

A was deposited after D [1]

(g) Large (white) crystals. [1]

Cools and crystallises slowly inside the Earth's crust. [1]

Total 9 marks

Use this page for your notes.

Answers to mock examination Paper 2

Question 1

(a)	Increases	[1]
(b)	81 g	[1]
(c)	34 g	[1]
(d)	47 g	[1]
(e)	Filter off the potassium dichromate crystals	[1]
	wash with a small volume of water.	[1]

EXAMINER'S TIP

The small volume of water is important. It removes soluble impurities. Use too much water and all the potassium dichromate will dissolve. This question is testing a quantitative approach to solubility. This is important at Higher tier.

Total 6 marks

Question 2

(a)	Flame test	[1]
	Orange flame with sodium, pinky-lilac flame with potassium (Both required for 1 mark)	[1]
(b)	Add dilute nitric acid and silver nitrate solution	[1]
	White precipitate with potassium chloride	[1]
	No precipitate with potassium sulphate	[1]
	Or	
	Add dilute hydrochloric acid and barium chloride solution	[1]
	No precipitate with potassium chloride	[1]
	White precipitate with potassium sulphate	[1]
(c)	Add sodium hydroxide solution	[1]
	Green precipitate with iron(II) sulphate	[1]
	Red-brown precipitate with iron(III) sulphate	[1]
(d)	Add sodium hydroxide solution and heat.	[1]
	Ammonium chloride produces ammonia gas that turns damp red litmus paper blue.	[1]
	No gas given off with magnesium chloride.	[1]

EXAMINER'S TIP

You should give the result with each chemical.

Total 11 marks

Question 3

(a)

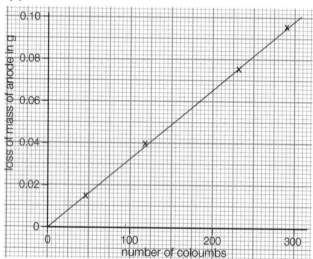

(One mark for correct scales and labelled axes, one mark for correct plotting, one for best line.) [3]

(b) (i)	190–195 coulombs	[1]
(ii)	Using the value of 192 coulombs.	
	0.064 g of copper deposited by 192 coulombs	[1]
	64 g of copper deposited by 192 000 coulombs	[1]
	1 mole of copper atoms deposited by 2 faradays.	[1]

EXAMINER'S TIP

1 mole of electrons is the same as 1 faraday. The examiner uses your value from (a) in the calculation.

(c)	$Cu(s) \rightarrow Cu^{2+}(aq) + 2e^-$	[4]

(One mark for correct formulae of reactants, one mark for correct formulae of products, one mark for balancing and one mark for state symbols.)

(d) Time = $\dfrac{\text{No. of coulombs}}{\text{Current}}$ [1]

$\dfrac{290}{0.2}$ [1]

$\dfrac{1450}{60}$ = 24 minutes (to the nearest minute) [1]

EXAMINER'S TIP

The first mark comes from the relationship. No. of coulombs = time (in seconds) × current (in A)

Total 14 marks

Question 4

(a) Atoms are closely packed [1]

Each atom surrounded by six in regular hexagon [1]

(b) Structure consists of free electrons [1]

Free electrons carry electric current [1]

(c)

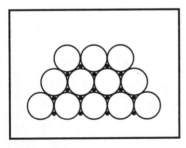

[1]

(d) In pure iron layers can slide over each other. [1]
In steel carbon atoms prevent atoms sliding. [1]

Total 7 marks

Question 5

(a) Dehydration/elimination [1]

(b) (i) $C_2H_4 + H_2O \rightarrow C_2H_5OH$ [1]
Any two from:

ethene and steam

passed over heated (phosphoric acid) catalyst

at high temperature and high pressure [2]

(ii) $C_{12}H_{22}O_{11} + H_2O \rightarrow 4C_2H_5OH + 4CO_2$ [3]
(One mark for correct formulae of reactants, one mark for correct formulae of products and one mark for balancing.)

Any two from:

sugar solution and yeast (or zymase)

temperature 35–40°C

absence of air or oxygen [2]

(iii) In USA and Western Europe there are plentiful supplies of ethene so this is a good source of ethanol. [1]

In less developed countries no ethene but plenty of sugar [1]

(c) (i) Isomers [1]

(ii) [2]

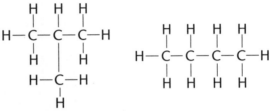

EXAMINER'S TIP

Draw out the structures fully as above. Check that they both have four carbon atoms and ten hydrogen atoms.

(iii) Bubble gases through bromine solution. [1]

Butene turns bromine from brown to colourless [1]

Butane has no effect on bromine [1]

Total 17 marks

Question 6

(a) Colourless [1]

to red [1]

(b) Bromide ions lose electrons to chlorine atoms [1]

Chlorine atoms gain electrons from bromide ions [1]

Bromide oxidised/chlorine reduced [1]

Total 5 marks

Additional information

Periodic Table

The Periodic Table

Key:

relative atomic mass	1.0
atomic symbol	**H**
atomic number	1
name	Hydrogen

The Periodic Table

1	2											3	4	5	6	7	0
																	4.0 **He** 2 Helium
6.9 **Li** 3 Lithium	9.0 **Be** 4 Beryllium											10.8 **B** 5 Boron	12.0 **C** 6 Carbon	14.0 **N** 7 Nitrogen	16.0 **O** 8 Oxygen	19.0 **F** 9 Fluorine	20.2 **Ne** 10 Neon
23.0 **Na** 11 Sodium	24.3 **Mg** 12 Magnesium											27.0 **Al** 13 Aluminium	28.1 **Si** 14 Silicon	31.0 **P** 15 Phosphorus	32.1 **S** 16 Sulphur	35.5 **Cl** 17 Chlorine	39.9 **Ar** 18 Argon
39.1 **K** 19 Potassium	40.1 **Ca** 20 Calcium	45.0 **Sc** 21 Scandium	47.9 **Ti** 22 Titanium	50.9 **V** 23 Vanadium	52.0 **Cr** 24 Chromium	54.9 **Mn** 25 Manganese	55.8 **Fe** 26 Iron	58.9 **Co** 27 Cobalt	58.7 **Ni** 28 Nickel	63.5 **Cu** 29 Copper	65.4 **Zn** 30 Zinc	69.7 **Ga** 31 Gallium	72.6 **Ge** 32 Germanium	74.9 **As** 33 Arsenic	79.0 **Se** 34 Selenium	79.9 **Br** 35 Bromine	83.8 **Kr** 36 Krypton
85.5 **Rb** 37 Rubidium	87.6 **Sr** 38 Strontium	88.9 **Y** 39 Yttrium	91.2 **Zr** 40 Zirconium	92.9 **Nb** 41 Niobium	95.9 **Mo** 42 Molybdenum	– **Tc** 43 Technetium	101 **Ru** 44 Ruthenium	103 **Rh** 45 Rhodium	106 **Pd** 46 Palladium	108 **Ag** 47 Silver	112 **Cd** 48 Cadmium	115 **In** 49 Indium	119 **Sn** 50 Tin	122 **Sb** 51 Antimony	128 **Te** 52 Tellurium	127 **I** 53 Iodine	131 **Xe** 54 Xenon
133 **Cs** 55 Caesium	137 **Ba** 56 Barium	139 **La** 57 Lanthanum	178 **Hf** 72 Hafnium	181 **Ta** 73 Tantalum	184 **W** 74 Tungsten	186 **Re** 75 Rhenium	190 **Os** 76 Osmium	192 **Ir** 77 Iridium	195 **Pt** 78 Platinum	197 **Au** 79 Gold	201 **Hg** 80 Mercury	204 **Tl** 81 Thallium	207 **Pb** 82 Lead	209 **Bi** 83 Bismuth	– **Po** 84 Polonium	– **At** 85 Astatine	– **Rn** 86 Radon
– **Fr** 87 Fracium	– **Ra** 88 Radium	– **Ac** 89 Actinium	– **Rf** 104 Rutherfordium	– **Db** 105 Dubnium	– **Sg** 106 Seaborgium	– **Bh** 107 Bohrium	– **Hs** 108 Hassium	– **Mt** 109 Meitnerium	– **Unn** 110 Unununilium	– **Uuu** 111 Unununium	– **Uub** 112 Unumbium		– **Uuq** 114 Ununquadium		– **Uuh** 116 Ununhexium		– **Uuo** 118 Ununoctium

lanthanides

140 **Ce** 58 Cerium	141 **Pr** 59 Praseodymium	144 **Nd** 60 Neodymium	– **Pm** 61 Promethium	150 **Sm** 62 Samarium	152 **Eu** 63 Europium	157 **Gd** 64 Gadolinium	159 **Tb** 65 Terbium	163 **Dy** 66 Dysprosium	165 **Ho** 67 Holmium	167 **Er** 68 Erbium	169 **Tm** 69 Thulium	173 **Yb** 70 Ytterbium	175 **Lu** 71 Lutetium

actinides

– **Th** 90 Thorium	– **Pa** 91 Protactinium	– **U** 92 Uranium	– **Np** 93 Neptunium	– **Pu** 94 Plutonium	– **Am** 95 Americium	– **Cm** 96 Curium	– **Bk** 97 Berkelium	– **Cf** 98 Californium	– **Es** 99 Einsteinium	– **Fm** 100 Fermium	– **Md** 101 Mendelevium	– **No** 102 Nobelium	– **Lr** 103 Lawrencium

Data sheet

Reactivity series of metals

Potassium   most reactive

Sodium

Calcium

Magnesium

Aluminium

Carbon

Zinc

Iron

Tin

Lead

Hydrogen

Copper

Silver

Gold

Platinum   least reactive

(Elements in italics, though non-metals, have been included for comparison.)

Formulae of some common ions

Positive ions		Negative ions	
Name	**Formula**	**Name**	**Formula**
Hydrogen	H^+	Chloride	Cl^-
Sodium	Na^+	Bromide	Br^-
Silver	Ag^+	Flouride	F^-
Potassium	K^+	Iodine	I^-
Lithium	Li^+	Hydroxide	OH^-
Ammonium	NH_4^+	Nitrate	NO_3^-
Barium	Ba^{2+}	Oxide	O^{2-}
Calcium	Ca^{2+}	Sulphide	S^{2-}
Copper(II)	Cu^{2+}	Sulphate	SO_4^{2-}
Magnesium	Mg^{2+}	Carbonate	CO_3^{2-}
Zinc	Zn^{2+}		
Lead	Pb^{2+}		
Iron(II)	Fe^{2+}		
Iron(III)	Fe^{3+}		
Aluminium	Al^{3+}		

Chemical equations

One of the most important skills you need to master for success in Chemistry at Grades B and above is the ability to write and balance equations.

The first stage in writing equations is probably to write a **word equation**. A word equation summarises a chemical reaction.

For example, sodium sulphate is formed when sodium hydroxide and sulphuric acid react. Water is also produced.

The word equation is:

sodium hydroxide + sulphuric acid → sodium sulphate + water

Having written the word equation, you should write the formulae of the reactants and products.

$$NaOH + H_2SO_4 \rightarrow Na_2SO_4 + H_2O$$

Now you have to **balance** the equation.

This means there must be the same number of each type of atom on each side of the equation.

You must not change any of the formulae.

$$2NaOH + H_2SO_4 \rightarrow Na_2SO_4 + 2H_2O$$

On each side of the equation there are two sodium atoms, four hydrogen atoms and six oxygen atoms.

The equation is balanced.

In marking an equation, the general rule for examiners is one mark for the correct formulae on the left-hand side, one mark for the formulae on the right-hand side and one for balancing.

This can be varied in individual cases. For example, if the formulae of the reactants are given or the equation needs no balancing.

Sometimes state symbols are used in equations to show the state of reactants and products.

These symbols are:

(s) solid

(l) liquid

(g) gas

(aq) aqueous solution (water) is the solvent.

So the equation of sodium hydroxide and sulphuric acid can be written as:

$$2NaOH(aq) + H_2SO_4(aq) \rightarrow Na_2SO_4(aq) + 2H_2O(l)$$

There is not usually a mark for state symbols unless they are expressly asked for.

Ionic equations

It is sometimes useful to write ionic equations.

For example, the reaction of sodium hydroxide and sulphuric acid can be expanded to show where ions exist.

$$2Na^+ + 2OH^- + 2H^+ + SO_4^{2-} \rightarrow 2Na^+ + SO_4^{2-} + 2H_2O$$

Ions can be crossed out when they appear unchanged on both sides. Finally the equation can be divided throughout by 2 to get it in its lowest form.

$$OH^- + H^+ \rightarrow H_2O$$

An ionic equation must have the same number of each type of atom on each side of the equation. Ionic equations are normally targeted at A or A*.

List of common equations

The following equations are common equations you might meet.
Make sure you can write these equations.

Sodium + chlorine → sodium chloride

$2Na(s) + Cl_2(g) → 2NaCl(s)$

Iron + chlorine → iron(III) chloride

$2Fe(s) + 3Cl_2(g) → 2FeCl_3(s)$

Chlorine + sodium bromide → bromine + sodium chloride

$Cl_2(g) + 2NaBr(aq) → Br_2(g) + 2NaCl(aq)$

Magnesium + sulphuric acid → magnesium sulphate + hydrogen

$Mg(s) + H_2SO_4(aq) → MgSO_4(aq) + H_2(g)$

Calcium carbonate + hydrochloric acid → calcium chloride + carbon dioxide + water

$CaCO_3(s) + 2HCl(aq)_2 → CaCl_2(aq) + CO_2(g) + H_2O(l)$

Sodium thiosulphate + hydrochloric acid → sodium chloride + sulphur dioxide + sulphur + water

$Na_2S_2O_3(aq) + 2HCl(aq) → 2NaCl(aq) + SO_2(g) + S(s) + H_2O(l)$

Hydrogen peroxide → water + oxygen

$2H_2O_2(aq) → 2H_2O(l) + O_2(g)$

Methane + oxygen → carbon dioxide + water

$CH_4(g) + 2O_2(g) → CO_2(g) + 2H_2O(l)$

Calcium hydroxide + carbon dioxide → calcium carbonate + water

$Ca(OH)_2(aq) + CO_2(g) → CaCO_3(s) + H_2O(l)$

Copper(II) carbonate → copper(II) oxide + carbon dioxide

$CuCO_3(s) → CuO(s) + CO_2(g)$

Calcium carbonate → calcium oxide + carbon dioxide

$CaCO_3(s) → CaO(s) + CO_2(g)$

Calcium oxide + water → calcium hydroxide

$CaO(s) + H_2O(l) → Ca(OH)_2(s)$

Copper(II) oxide + hydrogen → copper + water

$CuO(s) + H_2(g) → Cu(s) + H_2O(l)$

Iron + copper(II) sulphate → Iron(II) sulphate + copper

$Fe(s) + CuSO_4(aq) → FeSO_4(aq) + Cu(s)$

Carbon + oxygen → carbon dioxide

$C(s) + O_2(g) \rightarrow CO_2(g)$

Carbon dioxide + carbon → carbon monoxide

$CO_2(g) + C(s) \rightarrow 2CO(g)$

Carbon monoxide + iron(III) oxide → carbon dioxide + iron

$3CO(g) + Fe_2O_3(s) \rightarrow 3CO_2(g) + 2Fe(l)$

Calcium carbonate → calcium oxide + carbon dioxide

$CaCO_3(s) \rightarrow CaO(s) + CO_2(g)$

Calcium oxide + silicon dioxide → calcium silicate

$CaO(s) + SiO_2(s) \rightarrow CaSiO_3(l)$

Lithium + oxygen → lithium oxide

$4Li(s) + O_2(g) \rightarrow 2Li_2O(s)$

Sodium + water → sodium hydroxide + hydrogen

$2Na(s) + 2H_2O(l) \rightarrow 2NaOH(aq) + H_2(g)$

Nitrogen + hydrogen ⇌ ammonia

$N_2(g) + 3H_2(g) \rightleftharpoons 2NH_3(g)$

Ammonia + water → ammonium hydroxide

$NH_3(g) + H_2O(l) \rightarrow NH_4OH(aq)$

Ammonium hydroxide + nitric acid → ammonium nitrate + water

$NH_4OH(aq) + HNO_3(aq) \rightarrow NH_4OH(aq) + H_2O(l)$

Hydrochloric acid + sodium hydroxide → sodium chloride + water

$HCl(aq) + NaOH(aq) \rightarrow NaCl(aq) + H_2O(l)$

Hydrogen + oxygen → water

$2H_2(g) + O_2(g) \rightarrow 2H_2O(l)$

Word list

Success at GCSE Chemistry relies on using scientific language correctly. These exercises are to ensure you understand and use the language correctly.

Match these words to the definitions below.

alkali metal	atom	atomic number	compound	covalent
electron	giant structure	group	halogen	ion
ionic	isotope	mass number	molecular	neutron
noble gas	nucleus	period	proton	transition metals

1 The smallest particle of an element that can exist. ....

2 Vertical column in the Periodic Table. ....

3 An element in group 7 of the Periodic Table, e.g. chlorine, bromine. ....

4 An element in group 1 of the Periodic Table, e.g. sodium, potassium. ....

5 A substance formed by joining atoms of different elements together. ....

6 The block of metals between the two parts of the main block in the Periodic Table. ....

7 The number of protons in the nucleus of an atom. ....

8 A structure in which all of the particles are linked together by a network of bonds extending through the crystal. ....

9 Atoms with the same atomic number, but different atomic masses. ....

10 A horizontal row in the Periodic Table. ....

11 A positively charged particle in the nucleus of an atom. ....

12 An element in group 0 of the Periodic Table. ....

13 A negatively charged particle that orbits the nucleus of an atom and is responsible for electrical conduction in metals. ....

14 A positively or negatively charged particle formed when an atom or group of atoms loses or gains electrons. ....

15 The number of protons plus neutrons in an atom. ....

16 A type of bonding involving the sharing of one or more pairs of electrons. ....

17 A type of bonding involving complete transfer of one or more electrons from a metal atom to a non-metal atom. ....

18 The central part of an atom. ....

19 An unchanged nuclear particle similar in mass to a proton. ....

20 A type of structure built up of molecules. Substances with these structures have low melting points and boiling points. ....

Answers
1 atom; 2 group; 3 halogen; 4 alkali metal; 5 compound; 6 transition metals; 7 atomic number; 8 giant structure; 9 isotope; 10 period; 11 proton; 12 noble gas; 13 electron; 14 ion; 15 mass number; 16 covalent; 17 ionic; 18 nucleus; 19 neutron; 20 molecular.

Match these words to the definitions below.

alkane	alkene	cracking	crude oil	crust
erosion	extrusive rocks	flammable	fossils	fossil fuels
fractional distillation	fuel	hydrocarbon	igneous	intrusive rocks
magma	metamorphic	polymerisation	saturated compound	sedimentary
weathering				

1 The outer layer of the Earth. ....

2 A substance that burns and releases energy. ....

3 Describes a substance, e.g. petrol, which catches alight easily. ....

4 The remains of plant and animal bodies which have not decayed and disappeared but have been preserved. They may be found in sedimentary and metamorphic rocks but not in igneous rocks. ....

5 Type of rock that is composed of compacted fragments of older rocks which have been deposited in layers on the floor of a lake or sea, e.g. sandstone. ....

6 The process where rocks are broken down by chemical, physical or biological means. ....

7 Repetitive chemical combination of small molecules (called monomers) to form a large chain known as a polymer. ....

8 Semi-molten rock under the crust of the Earth. ....

9 A complex mixture produced by the action of high temperatures and high pressures on the remains of sea creatures in the absence of air. It is trapped below impermeable layers of rocks. ....

10 A compound containing carbon and hydrogen only. ....

11 A compound containing only single covalent bonds, e.g. tetrachloromethane CCl_4. ....

12 These include coal, oil and natural gas and are produced in the Earth over long periods of time. ....

13 The action of wind, rain, snow etc. on rocks. ....

14 A rock that crystallises on the surface of the Earth, e.g. basalt. ....

15 A family of hydrocarbons with a general formula of C_nH_{2n+2}. ....

16 A family of hydrocarbons with a general formula of C_nH_{2n}. ....

17 A method of separating a mixture of liquids using different boiling points. ....

18 Rocks that have cooled and crystallised from molten rocks. ....

19 A rock that has been altered by high temperatures and high pressures. ....

20 Igneous rocks that crystallise inside the Earth, e.g. granite. ....

21 A process where long chains are broken down into smaller units. ....

Answers

1 crust; 2 fuel; 3 flammable; 4 fossils; 5 sedimentary; 6 weathering; 7 polymerisation; 8 magma; 9 crude oil; 10 hydrocarbon; 11 saturated compound; 12 fossil fuels; 13 erosion; 14 extrusive rocks; 15 alkane; 16 alkene; 17 fractional distillation; 18 igneous rocks; 19 metamorphic; 20 intrusive rocks; 21 cracking.

Match these words to the definitions below.

acid   acid rain   alkali   alloy   anion
anode   aqueous solution   base   cathode   cation
displacement reaction   electrode   electrolysis   electrolyte   indicator
ion   neutralisation   oxidation   pH   reactivity series
redox reaction   reduction   salt

1   A substance that dissolves in water to form a solution with a pH value below 7.
....

2   The scale measuring acidity and alkalinity. ....

3   A substance that dissolves in water to form a solution with a pH value above 7.
....

4   The splitting up of an electrolyte, either molten or in aqueous solution, with electricity.
....

5   A list of metals put in order of their reactivity, with the most reactive at the top of the list.
....

6   A chemical that can show if a substance is an acid or an alkali by changing colour.
....

7   This is formed when a substance is dissolved in water. ....

8   Rain that contains above normal amounts of sulphur dioxide and oxides of nitrogen.
....

9   A substance formed as a product of neutralisation. ....

10   The conducting rod or plate which carries electricity in and out of an electrolyte during electrolysis. ....

11   A metal made by mixing two other metals together, e.g. brass from copper and zinc.
....

12   A negatively charged electrode in electrolysis. ....

13   A reaction in which a more reactive metal (or halogen) displaces (pushes out) a less reactive metal from a compound of the less reactive metal (halogen). ....

14   A reaction where both oxidation and reduction take place. ....

15   A reaction in which an acid reacts with a base or alkali. ....

16   The positively charged electrode during electrolysis. ....

17   A negatively charged ion, e.g. Cl^-, which moves towards the anode (positive electrode) during electrolysis. ....

18   A positively or negatively charged particle formed when an atom or group of atoms loses or gains electrons. ....

19   A reaction where a substance gains oxygen, loses hydrogen or loses electrons. The opposite of reduction. ....

20   A chemical compound which, in aqueous solution or when molten, conducts electricity and is split up by it. ....

21   A reaction where a substance gains hydrogen, loses oxygen or gains electrons.
....

22   A positively charged ion, e.g. H^+, which moves towards the cathode (negative electrode) during electrolysis. ....

23   A metal oxide which reacts with an acid to form a salt and water only. ....

Additional information

Match these words to the definitions below.

activation energy   ammonia   ammonium nitrate   catalyst
energy level diagram   enzyme   endothermic reaction   equilibrium
exothermic reaction   fermentation   formula mass   Haber process
nitric acid   reversible reaction

1 A substance that alters the rate of a reaction without being used up. ....

2 An acid produced by the oxidation of ammonia. ....

3 The process in which enzymes in yeast convert glucose into ethanol and carbon dioxide. ....

4 A colourless gas that turns red litmus blue. ....

5 A protein molecule that acts as a biological catalyst. ....

6 The energy required to start a reaction. ....

7 What we call a reversible reaction when the rate of the forward reaction equals the rate of the reverse reaction. ....

8 A reaction which can go forwards or backwards depending upon the conditions. ....

9 A reaction which gives out energy to the surroundings. ....

10 Mass in grams of the sum of the relative atomic masses. ....

11 A diagram showing the energy content at stages during a reaction. ....

12 A reaction that takes in energy from the surroundings. ....

13 Made by neutralising nitric acid and ammonia. ....

14 The name of the process involving the combination of nitrogen and hydrogen to make ammonia. ....

Answers

1 catalyst; 2 nitric acid; 3 fermentation; 4 ammonia; 5 enzyme; 6 activation energy; 7 equilibrium; 8 reversible reaction; 9 exothermic reaction; 10 formula mass; 11 energy level diagram; 12 endothermic reaction; 13 ammonium nitrate; 14 Haber process.

Index